EcoProduction

Environmental Issues in Logistics and Manufacturing

Series Editor

Paulina Golinska, Poznan, Poland

For further volumes:
http://www.springer.com/series/10152

About the Series

The EcoProduction Series is a forum for presenting emerging environmental issues in Logistics and Manufacturing. Its main objective is a multidisciplinary approach to link the scientific activities in various manufacturing and logistics fields with the sustainability research. It encompasses topical monographs and selected conference proceedings, authored or edited by leading experts as well as by promising young scientists. The Series aims to provide the impulse for new ideas by reporting on the state-of-the-art and motivating for the future development of sustainable manufacturing systems, environmentally conscious operations management and reverse or closed loop logistics.

It aims to bring together academic, industry and government personnel from various countries to present and discuss the challenges for implementation of sustainable policy in the field of production and logistics.

Subramanian Senthilkannan Muthu
Yi Li

Assessment of Environmental Impact by Grocery Shopping Bags

An Eco-Functional Approach

Springer

Subramanian Senthilkannan Muthu
SGS Hong Kong Limited
Global Sustainability Services
Hong Kong
Hong Kong SAR

Yi Li
Institute of Textiles and Clothing
Hong Kong Polytechnic University
Hong Kong
Hong Kong SAR

ISSN 2193-4614 ISSN 2193-4622 (electronic)
ISBN 978-981-10-1331-7 ISBN 978-981-4560-20-7 (eBook)
DOI 10.1007/978-981-4560-20-7
Springer Singapore Heidelberg New York Dordrecht London

Printed on acid-free paper

Springer is part of Springer Science+Business Media (www.springer.com)

Preface

Eco-functional assessment is a pretty new concept, developed by us to evaluate any consumer product belonging to different categories such as textiles, plastics and paper. A product commences its life cycle from raw material extraction and spans through other phases such as manufacturing, consumer use and finally ends its life at end-of-life phase. Evaluation of the environmental impacts of a product in its entire life span is of pivotal importance in order to to reduce them and this can be deciphered by the life cycle assessment technique. Life Cycle Assessment (LCA) is one of the widely used and well-known concepts to evaluate the potential environmental impacts of a product, process or service. LCA can be carried out in many stages and some of the most familiar stages include cradle to gate and cradle to grave.

Influence of functionality in each life cycle phase of a product and consequently the life cycle impacts of the product under question is of significant importance. Similarly, the influence of consumer behaviour in deciding the life cycle impacts is crucial as well and has to be included in assessing the environmental impacts of a product. Current LCA methods and models do not aid a way to include these aspects, which will be fulfilled by our research entitled, "Eco-Functional Assessment". In this assessment, any product on earth can be evaluated in terms of functional and ecological aspects and consumer behaviour in a single platform. We did develop the basic principle and theoretical framework of eco-functional assessment. Based on the theoretical framework, an eco-functional model was developed. This model comprises of four inputs namely raw materials, process of manufacture, functional and ecological properties and five outputs namely quality, functionality, Reduce, Reuse and Recycle (3R's), environmental and human impacts. Any environmental impact from the long list of impacts detailed in LCA models can be studied such as carbon footprint, ecological footprint and environmental damage. The toughest task was to connect these inputs and outputs, which we did by using certain rules, sub models, equations and formulae.

Shopping bags are one of the essential entities in our daily life and their environmental impacts are indispensable. Every individual and every country is keen on this topic and every one is aware of the potential impacts created by a shopping bag. Thanks to media and many governments who have already implemented ban on use of plastic bags, through which even common public is aware of the environmental impacts of shopping bags to some extent and they are encouraged to bring their own bags for grocery shopping. However, there is no proper literature in book format or even in a monograph format, dedicated only for shopping bags talking about their different types, manufacturing processes, their environmental impacts. Hence we decided to write this book and this book, first of its kind, will fulfil this need and bridge this literature gap.

Once we decided the framework of eco-functional model, we were thinking of selecting a suitable product to demonstrate the applications of this framework. We started this research in 2008, during that time and even now shopping bags used for grocery shopping is a debatable a topic in terms of their eco-impacts. In 2008, China government has banned HDPE bags and started levying LDPE bags. In 2009 July HKSAR government has started introducing the levy on plastic shopping bags and people were encouraged to go for alternative reusable bags and they were forced to bring their own bags. Understanding this scenario, we decided to utilise different shopping bags used for grocery shopping bags as a base product to demonstrate our framework and the eco-functional model. This certainly was a correct choice to test the suitability and robustness of our framework and model to test different kinds of materials and products made out of different technologies. We did choose shopping bags made out of plastic (HDPE, LDPE), paper, polypropylene nonwoven bags, polyester nonwoven bags and cotton woven bags for our assessment. This list formed the wide array of materials and different technologies to evaluate the concept and these bags were categorised into two main groups namely single use and reusable bags.

Influence of consumer behaviour is the key in deciding the environmental impacts of any product, especially for products such a shopping bags, which portray the symbol of throw-away society. So far, different LCA studies conducted on shopping bags included an assumption to model the use and disposal phases, which are certainly very far from reality. Hence, we decided to study the consumer behaviour by means of administering a questionnaire survey in China, Hong Kong and India and derived the values pertaining to use and disposal of shopping bags amongst various user groups. These values can directly be plugged in LCA calculations and we did employ these values in our eco-functional model. Likewise, influence of functionality is an important element to be included in LCA calculations and this was also done in our assessment.

This book is the first book on earth dealing with the functionality part of shopping bags. The parameters of functionality for shopping bags and the test results of different grocery shopping bags in the light of functional parameters are well discussed. Similarly, this book covers different aspects in the entire life cycle phases of shopping bags starting from cradle to grave stages. This book reviewed the novel models developed by us to evaluate different raw materials in terms of environmental impact and ecological sustainability. We did develop a concept that defines recyclability and a model to quantify the recyclability of different raw materials in terms of their environmental and economic gains. This concept and model is also briefed in this book.

There are no scientific equipments and methods available yet to test the reusability, holding capacity and impact strength of different shopping bags. We did develop a novel tester named, "Eco-Functional Tester" to quantify these parameters of different shopping bag. This tester and the testing results from this equipment are briefly discussed. Results from these different models and this tester were included in our eco-functional assessment model.

Polypropylene nonwoven bags are gaining popularity since the introduction of ban on plastic bags. We did study the life cycle inventory of polypropylene nonwoven bags from a nonwoven bags manufacturing factory and quantify the life cycle impacts of these bags. We also found out the hot-spots in the manufacturing processes of polypropylene nonwoven bags. We did attempt an innovative way to quantify the life cycle impacts of different grocery shopping bags pertaining to the consumers living in China. For this, well-known LCA equations to characterise the impacts were used along with the characterisation and normalisation factors for China. These innovative pieces of work done by us are included in this book.

All these innovative aspects are included in our eco-assessment framework and the eco-functional model and this book deals with all these in a comprehensive way. This book can elaborate the details of entire life cycle phases of shopping bags and the functional aspects of shopping bags, along with the detailed concept of eco-functional assessment methodology. This book can be a very good reference for students and researchers.

Contents

Chapter 1
Basic Introduction to Shopping Bags and Eco-Functional Assessment of Shopping Bags

1.1 Introduction

No product's life cycle starts and ends without leaving its impact on our living planet. Manufacturing, consumption and disposal of every product create environmental impacts. Degree of environmental impacts varies from product to product and it is largely influenced by many factors, among them the useful life of the product deserves significant importance. In general, products with short span of life create more impacts than the ones which last longer. Again, this cannot be generalized, but as a matter of fact, functional life is the key to decide the eco-impacts of any product. Assessment of functional life coupled with its life cycle impacts is defined by us as, "Eco-functional Assessment". The concept of "Eco-functional assessment" is important for all products and it finds its suitability of application for all products produced on earth; however it is crucial for products with short span of life to be assessed in terms of their eco-functionality. One of the key products regarded as the symbol of throw away society which has very short span of life and has to be assessed in terms of eco-functionality is shopping bags used for grocery purposes.

Shopping bags are one of the essential items of one's life these days and it has become so inevitable with the change in shopping cultures compared to the previous decades. These days, one cannot live without supermarkets and hence the shopping bags. Super and hyper markets are found in every corner of any city now-a-days and every market uses grocery shopping bags. There are many different variety of shopping bags used for grocery purposes worldwide. In this long list, plastic bags, paper bags, reusable bags made out of plastic and cotton are the most widely used and they top the entire list. Shopping bags can be categorized into two types generally as single use and reusable ones. When it comes to shopping bags, single use plastic bags flicker into every one's mind and they represent a huge quantity and take over their rivals.

Life cycle of shopping bags starts from raw material extraction followed by other phases as shown in Fig. 1.1 and ends at the disposal phase. Every phase needs deployment of a lot of resources such as energy, chemicals, other accessories and water and each phase of life cycle is responsible for the creation of

S. S. Muthu and Y. Li, *Assessment of Environmental Impact by Grocery Shopping Bags*, EcoProduction, DOI: 10.1007/978-981-4560-20-7_1,

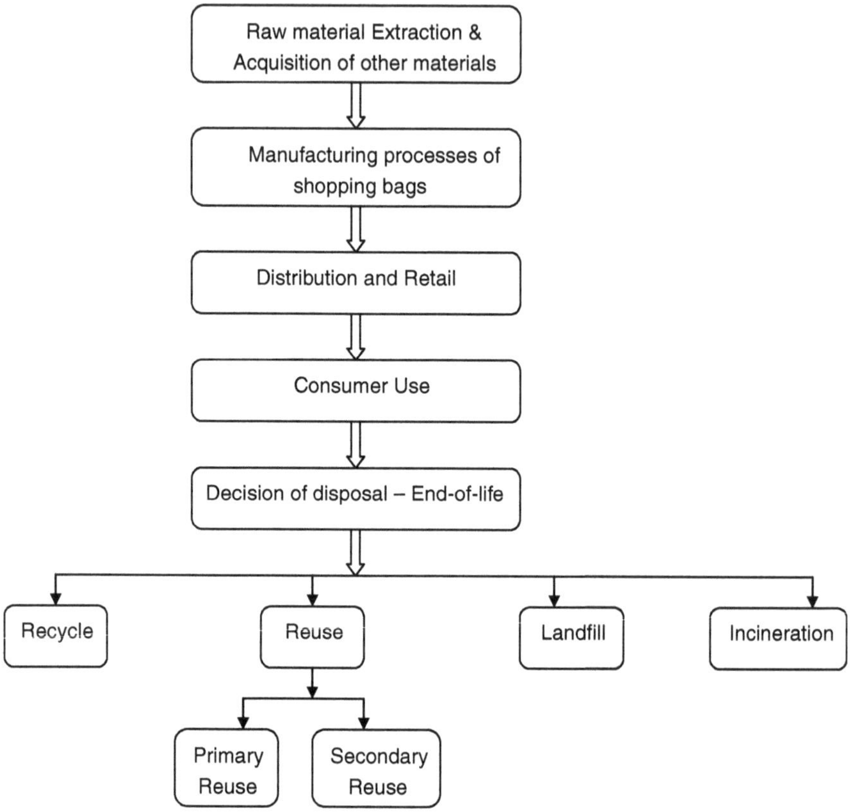

Fig. 1.1 Life cycle model of grocery shopping bags

multiple life cycle impacts. These impacts go to peak when the life of the product under question is very short, since an immediate new product has to replace the current product after its life ends. And the very important thing is most of such short shelf life products especially made out of plastic pose huge environmental threats at the disposal phase. This is why single use shopping bags, which are perceived to be a symbol of throw-away society, are given serious consideration in order to reduce the eco-impacts made by them. It does not mean that the reusable bags made out of either plastic (polypropylene) or cotton is completely environmental friendly either and they create no environmental impacts in their entire life cycle. They also create eco-impacts, however, they try to alleviate the impacts to certain level by means of being reused many more times till disposed, when compared to single use grocery bags. Reusable bags compensate the environmental impacts by means of their reusability. If the reusable bags are thrown after the first use, their life cycle impacts will be very higher than the single use ones. The entire crux of life cycle assessment is to assess the impacts in the whole life cycle.

This chapter discusses about the different types of shopping bags in detail along with the eco-impacts of shopping bags and the functionality of shopping bags used for grocery purposes. This chapter also deals with the concept of eco-functional assessment in detail and how it can be applied for grocery shopping bags.

1.2 Shopping Bags-Different Types and Purposes

Web based dictionaries define shopping bags as follows. Since there are many dictionaries available on-line today, a sample of three definitions is given below:

- A shopping bag is a strong container with one or two handles, used to carry things in when you go shopping [1];
- A bag made of plastic or strong paper (often with handles); used to transport goods after shopping [2];
- A bag (as of strong paper) that has handles and is intended for carrying purchases [3].

All the three definitions given above (and even in any dictionary for that matter) specify two main identities of shopping bag namely a bag with handles and intended for carrying purposes. Shopping bag serves all the functions of packaging namely contain and protect the product [4], where the functional aspects of the shopping bags play a vital role. Apart from the functionality aspects, shopping bags are also intended to be used for promotional purposes, showcase as a status symbol and for branding as well [4–6].

Shopping bags can be classified into two main types based on the purpose namely grocery shopping bags and non-grocery shopping bags. Size of the bag which decides the volume is a critical parameter to determine this distinction between grocery and non-grocery bags. Grocery shopping bags are generally in medium size in the range of 2.5–5 gallons in volume and for non-grocery purposes, much larger volumes are used [7]. This entire book revolves around the shopping bags used for grocery purposes.

Grocery shopping bags are primarily classified in terms of the number of times they can be used. This classification is very important, since it is the base to decide the eco-impacts made by different shopping bags. Depending on the number of times a bag can be used, shopping bags are classified into single use bags (also termed as disposable bags) and reusable bags.

Grocery shopping bags are also classified by means of the material content or material type. As per this classification, grocery bags are grouped into four categories namely plastic bags, paper bags, textile woven bags, reusable nonwoven bags. This classification also separates different bags according to the technology used to manufacture them.

Polyethylene in low density (LDPE), linear low density (LLDPE) and high density forms are used to produce plastic bags. As the name implies, paper bags are made out of paper material. Textile woven bags are made out of cotton essentially, produced by weaving technology. Other materials such as polypropylene are also used to produce woven bags. Nonwoven bags are made out of non-woven technology non-woven by employing polypropylene or polyester essentially to make reusable bags. Polypropylene nonwoven bags are very common in reusable category.

1.3 Eco-Impacts of Shopping Bags

Environmental impacts created by any product is of huge concern these days. All products produced on earth have environmental risk. The concern becomes very serious when a product has a short span of life, such as grocery shopping bags. Environmental damage created by grocery shopping bags is a well known issue and also a widely discussed topic by media, governments, NGO's, researchers and even consumers.

All types of grocery bags create environmental impacts, however with varied magnitude and point of time. Life cycle environmental impacts created by single use bags are critical compared to reusable ones. Reusable ones try to compensate impacts created during their manufacturing phase by being reused many times. However, due to the functionality issue or customer attitude if the reusable bags are thrown away before their desired number of use, the magnitude and severity of environmental concerns are really huge and they even surpass the impacts of single use bags.

Plastic, paper, cotton, linen are the widely used raw materials to produce grocery bags of different types using various bag manufacturing techniques. Extraction of raw materials, manufacturing processes of bags, transportation both internally within the factory and externally from the factory to ware house, distribution, retail and finally disposal stage of bags create multiple environmental impacts. Depending upon the raw material and manufacturing techniques, life cycle impacts will vary amongst different types of grocery bags. Use and disposal phases are critical in deciding the life cycle impacts. The disposal options of bags also determine the degree of magnitude of impacts; this issue is further influenced by the biodegradation of various raw materials employed for grocery bags, if they are opt to be disposed at landfills.

Eco-impacts of plastic grocery bags are well known to everyone including end customers. One of the main reasons behind this is, plastic bags being inexpensive and given by almost every supermarket for free. Marketing and promotional interests are also built into this. When this issue has become so big after every country realized the potential issues pertaining to plastic grocery bags, environmental levy has been introduced. This levy has been introduced in almost all countries now and this was implemented in different countries at various points of time. Followed by this issue, reusable bags have become more familiar; consequently to address the eco-impacts created by single use bags. Eco-impact of various grocery bags will be discussed in necessary depth in the forthcoming chapters.

1.4 Functionality of Shopping Bags

Functionality means the functional attributes or properties of products, which decide their ultimate useful life time of them. For any product, functionality is the key, since it is directly linked to the life of the product. Different products have their own attributes, which decide their functionality and consequently the life of those products. Life of the products is decided by the combination of functional attributes and the consumer's attitude. However, at many instances, the influence of functionality surpasses the consumer's attitude and moreover, functionality is one of the key aspects which govern the consumer's attitude towards disposing a product.

Shopping bags used for grocery purposes have their own functional attributes. They have a long list of functional properties beginning from material composition and extend to areal density, thickness, various strength related properties, permeability attributes, color fastness tests and safety. This list also includes various safety parameters such as pH, formaldehyde, heavy metals, azo dyes. This comprehensive list consists of basic functional parameters; however, certain special parameters such as impact strength and load bearing capacity of bags also have to be added in the list to decide the functionality of the shopping bags.

1.5 Eco-Functional Assessment of Shopping Bags

Another property which assumes equal significance as functional properties is ecological properties, which relates to the life cycle of products under question. Ecological properties attempt to trace the entire life cycle of products from the beginning till the end of life cycle chain. These properties deserve emergent attention due to the alarming environmental issues surrounding us. Similar to functional properties, there is a long list of aspects which govern the ecological properties of a particular product.

Amongst various parameters that decide ecological attributes of shopping bags, 3R's (reduce, reuse, recycle) and biodegradation deserve special attention. It's pivotal to combine functional and ecological properties together in a single platform to decide the environmental impacts made by any product, which includes shopping bags as well. There are no attempts made so far and reported in the literature to amalgamate these properties under one roof. It is worthwhile to combine these two properties of shopping bags, which is the heart of this "Eco-functional Assessment research." The reason behind combining of these two properties together is that they go hand in hand in all circumstances. They are interrelated to each other in the sense that the functional properties govern the ecological properties. Both can be merged with each other in various ways. For instance, if a product is bestowed with better functional attributes, it will eventually last long and adjourns the arrival of a new product and its associated life cycle impacts. This book will enumerate such links in detail and explore the concept of eco-functional assessment with necessary depth and breadth.

1.6 Concluding Remarks

This introductory chapter highlighted the eco-impacts of shopping bags used for grocery purposes. General life cycle model of shopping bags was discussed in this chapter. Various shopping bag types and their purposes were introduced briefly. Functional and ecological properties of shopping bags were also discussed. The importance of eco-functional assessment and its rationale were also briefed in this chapter.

References

1. http://www.collinsdictionary.com/dictionary/english/shopping-bag
2. Thesaurus. http://www.thefreedictionary.com/shopping+bag
3. http://www.merriam-webster.com/dictionary/shopping%20bag
4. Kotler P, Armstrong G (2010) Principles of marketing, 13th edn. Prentice-Hall, Upper Saddle River
5. Sayles J (1996) "Introduction", shopping bag design 2. Rockport publishers, OH, distributed by North Light Books
6. Prendergast G, Ng SW, Leung LL (2001) Consumer perceptions of shopping bags. Mark Intell Plan 19(7):475–482
7. Wikipedia. http://en.wikipedia.org/wiki/Shopping_bag

Chapter 2
Manufacturing Processes of Grocery Shopping Bags

2.1 Introduction

For the products which do not last long or the products which do not create impacts in the use phase in their total life cycle, the manufacturing sequence or the process of manufacture would be a chunk in contributing towards major environmental impacts. This analogy holds good especially for grocery shopping bags, whose shelf life is pretty small and almost, do not have use phase impacts except for woven shopping bags, which do require washing between intermittent uses. Four major variety of shopping bags are mainly used for grocery purposes namely plastic, paper, nonwoven and woven shopping bags. This chapter deals with the manufacturing processes of these four types of shopping bags used for grocery shopping.

2.2 Plastic Bags

Plastic bags occupy a significant proportion of grocery shopping bags. Plastic bags are preferred mostly for grocery shopping, since they are more convenient to carry by customers for grocery purposes and also are inexpensive. Plastic bags are more appealing to consumers and retailers owing to these two factors [1].

Out of many types of polyethylene, high density polyethylene (HDPE) is used significantly for manufacturing plastic bags. Majority of the plastic grocery bags are produced by HDPE. There are two more variants of polyethylene used to produce plastic grocery bags apart from HDPE, which are low-density polyethylene (LDPE) and linear low-density polyethylene (LLDPE). Glossy bags used in shopping malls as carry bags are made out of LLDPE and a filmy type of very thin bags are made out of LDPE [1, 2].

Plastic bags, made out of any of the three types discussed above, are made from non-renewable resources, where the key ingredients are petroleum and natural gas [2]. Plastic is obtained as a by-product from the oil refining process

S. S. Muthu and Y. Li, *Assessment of Environmental Impact by Grocery Shopping Bags*, EcoProduction, DOI: 10.1007/978-981-4560-20-7_2,

and the oil used for manufacturing plastic bags accounts to 4 % of the world's total oil production [3]. Detailed production steps of plastic bags manufacturing process is outlined in Fig. 2.1.

Plastic bag manufacturing process starts from oil cracking process followed by many other processes to obtain polyethylene. Resin form of polyethylene is utilized to manufacture plastic bags by blown film extrusion, which is also termed as, "Tubular film process". While processing polyethylene to manufacture plastic bags, many additives such as anti-block (prevents the sticking of plastic bag layers), slip (enables the plastic bag to open easier) and Ultra Violet Inhibitor (UVI to prevent the weakening of strength and fading of colour by UV rays) are added [4].

As depicted in Fig. 2.1, polyethylene is fed into the hopper and the hopper in turn transfers the resin feed into the extruder. Resin is being melted with the aid of heating constituents and the turning movement of extruder screw. Molten form of resin is forced through the extruder and flows up and over the circular die in a uniform fashion. Then it passes through the bag manufacturing elements of blow moulding part as described in Fig. 2.1 [4]. An air-filled plastic bubble is created once the air is blown through the plastic, which is responsible to move the plastic

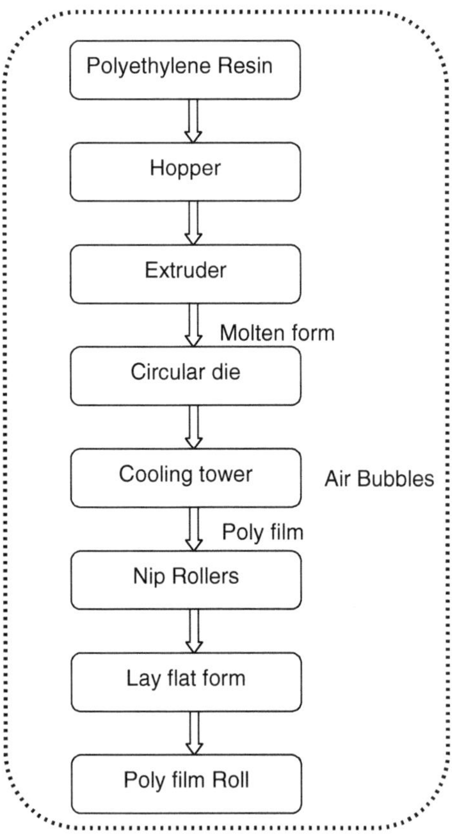

Fig. 2.1 Plastic bags manufacturing process

upward into a tube-shaped die. The control exercised by the manufacturers over the blown extrusion process help them to get the desired size and thickness of plastic bags in terms of shaping a plastic bag [5]. Finally, the produced plastic tube enters into a bag making machine directly or with the aid of rollers. This machine heat seals one end of the tube together to prepare the bottom part of a bag, at the same time a cutter at the other end does a precise form of opening [5].

2.3 Paper Bags

Paper bags are produced out of a renewable resources, vis-à-vis plastic bags. Pulpwood from trees is utilized to produce paper grocery bags. Paper bags manufacturing process constitutes three phases namely [6]:

- Preparation of paper
- Bag preparation process
- Bags assembly process.

These three phases and different processes involved in each phase are diagrammatically presented in Fig. 2.2.

In the first phase, after the trees are cut down, they are transported to a paper industry. Bark removal is the first process in a paper mill, followed by cutting the trees into small pieces. Those cut pieces will be treated with chemicals to remove the cellulose fibres from the remaining wood. Cellulose fibres will then be washed, bleached and dried before being rolled into flat sheets.

In the second phase, as shown in Fig. 2.2, paper bags will be applied with a coating by spray application followed by preparing the bags to the desired shape with printing of company name. Third phase comprises of activities related to assembling of bags, as indicated in Fig. 2.2 [6].

2.4 Nonwoven Bags

The third category of grocery bags is nonwoven bags which come under the category of reusable bags. Non-woven bags can be made from various raw materials namely polypropylene and polyester. Nonwoven grocery bags made from Polypropylene (PP) are the most commonly used in the market for grocery shopping.

Similar to plastic bags made out of HDPE/LDPE, nonwoven bags manufacturing process also starts with the oil refining process to obtain the raw material, i.e. polypropylene. The entire process of producing a PP grocery bag constitutes three steps. The first step is the production of PP, which is derived from non-renewable resources where the key ingredients used to produce it are petroleum

Fig. 2.2 Paper manufacturing process [6]

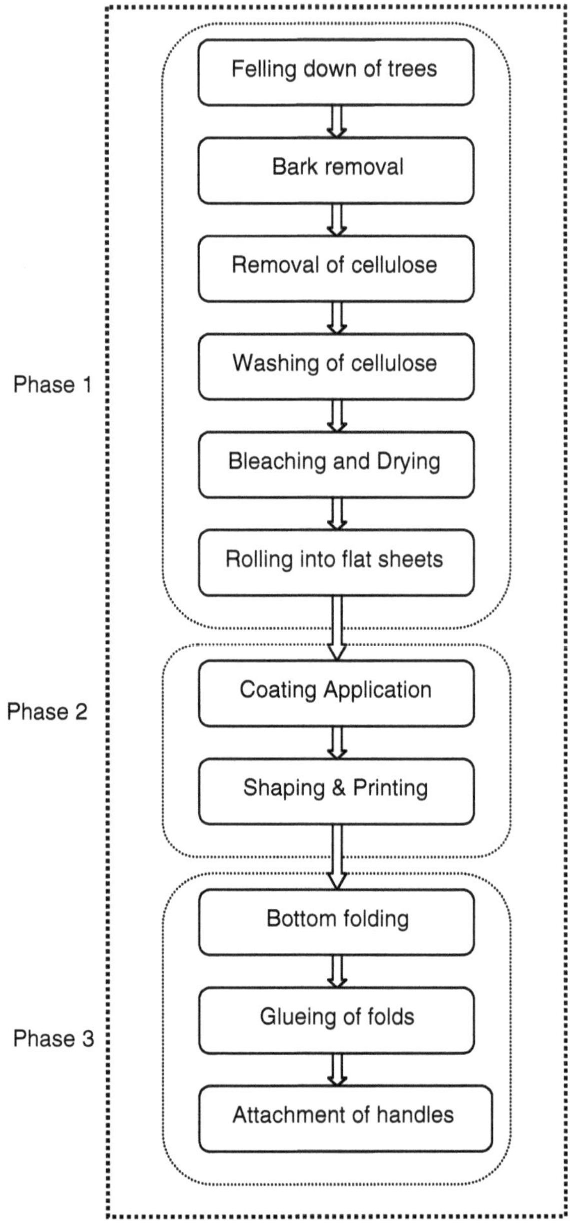

and natural gas. PP is produced by the polymerization process of propylene. Propylene, again, is a gaseous by-product obtained during petroleum refining. This production process happens in the presence of a catalyst under carefully controlled heat and pressure [7].

The second step is the spun bonding process, where spun bonded nonwoven fabrics are produced from PP resin. Spun bonding is a one step process of producing fabrics from plastic resin and it one of the commercially available polymer-laid processes [7, 8]. Different processes involved in spun bonding include polymer melting, filtering and extrusion, quenching, drawing, lay down on forming screen followed by the bonding process and finally the produced spun bonded fabric is rolled up.

The next stage constitutes different processes namely cutting, screen printing and sewing. These stages are outlined in Fig. 2.3.

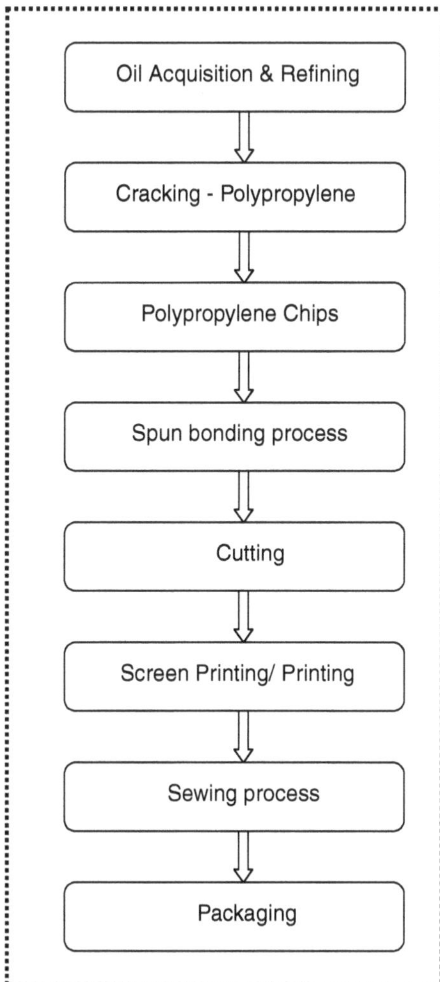

Fig. 2.3 Outline of nonwoven bags manufacturing process

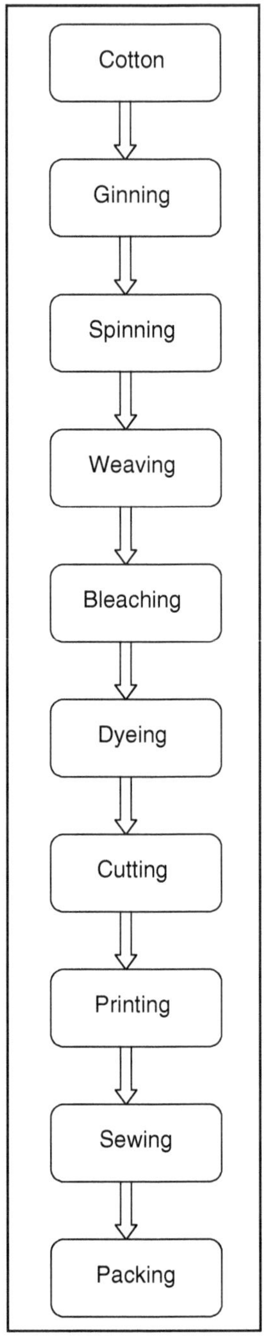

Fig. 2.4 Outline of woven (*cotton*) bags manufacturing process

2.5 Woven Bags

The next category of reusable bags is woven bags, which are named by the process of manufacture, i.e. weaving process. Woven bags are produced by both renewable and nonrenewable source of raw materials. Polypropylene is also employed to produce woven reusable bags. Very common type of woven bags is made by natural textile fibres such as cotton and hemp.

The production process of a woven bag is lengthy and involves a longer supply chain compared to a non-woven bag. For cotton woven bags, the manufacturing sequence starts from growing of cotton (either conventional or organic cotton), followed by separation of cotton fibres from seed cotton. Further, cotton fibres are spun into a yarn by spinning process and the yarns are converted into fabrics by weaving process. Fabrics are further processed by chemical processes such as bleaching and dyeing.

Further to which processes of garmenting such as cutting, screen printing and sewing will be completed to produce a woven bag. Production process of woven bags is outlined in Fig. 2.4. The outline remains same for other natural textile fibres such as hemp employed to manufacture woven bags.

2.6 Concluding Remarks

Though there is plenty of shopping bag categories available, plastic, paper, non-woven and woven bags top the list and are widely used for grocery shopping. Plastic and paper bags belong to single use bags category. The nonwoven bags made out of polypropylene and polyester, cotton woven bags are the important ones in the reusable bags category. In the life cycle of shopping bags, manufacturing processes play a major role in terms of occupying significant life cycle impacts. This is very much true especially in case of single use bags category. Each bag has its unique production process and method. In this chapter, manufacturing processes of different shopping bags used for grocery purposes were discussed in detail.

References

1. Ellis S, Kantner S, Saab A, Watson M (2005) Plastic grocery bags: the ecological footprint. Available at: http://www.vipirg.ca/archive/publications/pubs/student_papers/05_ecofootprint_plastic_bags.pdf. Accessed on Dec 2005
2. Lajeunesse S (2004) Plastic bags. Chem Eng News 82(38):51
3. Decker R, Graff A (2013) Paper vs. plastic bags? Can be viewed at: http://www.angelfire.com/wi/PaperVsPlastic/. Accessed on 02 Mar 2013

4. McWilliams JT (2013) 10 steps to plastic bags. Available at: http://www.5starwriting.com/wp-content/uploads/2011/06/10_Steps_To_Plastic_Bags_May06.pdf. Accessed on 27 Mar 2013
5. Norman W (2013) Procedure for plastic bag manufacturing. Available at: http://www.ehow.com/facts_6819404_procedure-plastic-bag-manufacturing.html. Accessed on 27 Mar 2013
6. David I (2013) Process for making paper bags. Available at: http://www.ehow.com/how-does_4683110_process-making-paper-bags.html. Accessed on 27 Mar 2013
7. Maier C, Calafut T (1998) Polypropylene—the definitive user's guide and data book. William Andrew Publishing, New York
8. Bhat GS, Malkan SR (2007) Polymer-laid web formation. In: Russell SJ (ed) Handbook of nonwovens. Woodhead Publishing, Cambridge, UK

Chapter 3
Life Cycle Assessment of Grocery Shopping Bags

3.1 Introduction

Any product produced on earth, including shopping bags begin their journey after depletion of enormous resources and its associated environmental impacts. From the beginning of life cycle till the product is disposed, every product is responsible to create many vulnerable impacts on our living planet. Degree of impacts vary between different products depending on many factors such as type of raw materials used, employment of renewable or non-renewable materials to produce them, length of supply chain link involved in the entire production process, energy requirement, demand of different consumables, and very importantly the length of life. These factors influence the environmental impacts created by various grocery bags namely plastic, paper, nonwoven and woven bags. Magnitude of environmental impacts is different between single and reusable shopping bags, due to the difference in the points discussed above. This chapter reviews the concept of life cycle assessment and the study of life cycle assessment of various shopping bags used for grocery shopping.

3.2 Concept of Life Cycle Assessment

Quatifying the environmental impacts produced by various products produced on earth is essential to reduce those impacts. Among the different techniques used to study the environmental impacts created by a product, life cycle assessment (LCA) is one of the most widely used and popular techniques. LCA examines the product from the initial (cradle) to the final stage (grave) and quantifies the environmental impacts created by a product in its entire life cycle.

Since the detailed explanation of life cycle assessment is out of the scope of this book, this chapter briefly discusses the concept of life cycle assessment. A life-cycle assessment (LCA) is an analytical tool which helps us to understand the environmental impacts from the acquisition of raw materials to final disposal [1].

S. S. Muthu and Y. Li, *Assessment of Environmental Impact by Grocery Shopping Bags*, EcoProduction, DOI: 10.1007/978-981-4560-20-7_3,

As defined by The Society of Environmental Toxicology and Chemistry (SETAC), LCA is an iterative process being employed to evaluate the environmental burdens associated with a product, process or activity. It works by identifying and quantifying the energy and materials used and the wastes released to the environment, in order to assess their environmental impact and identify and evaluate opportunities to effect environmental improvements. The assessment includes the entire life cycle of the product, process or activity, encompassing extracting and processing raw materials; manufacturing, transportation and distribution; use, reuse, maintenance, recycling and final disposal [2].

A product's life cycle begins from the extraction of the potential raw materials for its manufacture, followed by the manufacturing process of the product in question with the aid of energy, chemicals, water and other required inputs. Manufacturing stage is followed by transportation to the consumer, the use phase and finally the life cycle of a product ends at disposal stage. Life cycle assessment, includes all these stages in order to trace the impacts created by the product over its lifetime.

Life cycle assessment can be performed in different stages such as cradle to gate or cradle to grave or cradle to cradle stages. Cradle to gate is a partial life cycle study, covers the impacts from cradle to the production gate (distribution, use and disposal stages are excluded). Cradle to grave is a full life cycle study which includes all life cycle stages till disposal stage. Cradle to cradle is an explicit category of LCA, where the disposal stages involve recycling process from which an identical raw material is produced and the process becomes a closed loop.

According to the standards earmarked by ISO to conduct LCA, i.e. ISO 14040 and ISO 14044, an LCA study essentially consists of four interconnected steps/phases given below [3, 4]:

- Goal and scope definition
- Inventory analysis
- Impact assessment
- Interpretation.

In the first step (goal and scope definition), the term 'goal' is used to specify the application of the study, to state the very purpose of pursuing the study and also to identify the target audience. The 'scope' prescribes the breadth, the depth and complete details of the study. It is vital to define a functional unit as the object of the life cycle assessment study and the boundaries of the system under investigation with clear specifications for data quality requirements. This step and the following inventory analysis correspond to ISO 14040 and ISO 14044 [3, 4].

The second step, Inventory Analysis, (LCI—Life Cycle Inventory) focuses on analyzing the different flows of material and energy corresponding to the production process and the environment. Input flows refer to the various resources like raw materials, energy, land or indeed any factor in the production process. Output flows refer to any sort of emissions to air, water or to land.

The next step—Impact Assessment (LCIA—Life Cycle Impact Assessment), explores the results of the inventory analysis in terms of the environmental impact.

The effects identified in this step can be compared to create an overall assessment of the products under investigation. In the impact assessment phase, LCIA consists of both obligatory and optional elements in accordance with ISO 14040. In brief, this step consists of selecting and defining impact categories such as global warming, acidification, eutrophication, human toxicity, ozone depletion, photo-oxidant formation, depletion of abiotic resources and aquatic and terrestrial toxicity measures, and classifying them by assigning the results of the Impact Assessment to the relevant impact categories. The final step in the LCA process is interpretation, which is in accordance with ISO 14040 and ISO 14044 [3, 4] and aims to draw conclusions from the study and make suitable recommendations to mitigate the major impacts encountered.

3.3 Environmental Impact Assessment of Raw Materials used for the Production of Shopping Bags

This section deals with an unique model developed by authors to quantify the environmental impact made by various raw materials used for shopping bags and to position them in terms of ecological sustainability. This model was developed to evaluate a wide range of textile raw materials and also other raw materials used for the production of a popular variety of shopping bags by considering the major contributing factors in terms of environmental impact during the manufacturing phase (starting from growth /extraction stage to production of a useful fibre which can be spun). Consideration of environmental impact and ecological sustainability involved analysis of the following factors:

- The amount of oxygen produced/ carbon dioxide absorbed consequently contributing to off-set global warming during the production phase of a fibre;
- Utilization of renewable resources;
- Usage of land;
- Usage of fertilizers and pesticides;
- Utilization of renewable resources;
- Consumption of energy, water in the production phase;
- Amount of green house gases emitted during production;
- Fibre/raw material recyclability;
- Biodegradability of the material under question.

This model was developed by Muthu et al., in 2012 [5] to fill the gap of no unique model availability to evaluate the different textile raw materials and other raw materials used for shopping bags in terms of their environmental impact and ecological sustainability. This assessment included a long list of major textile fibres such as conventional cotton, organic cotton, hemp, polyester, nylon and other raw materials, such as paper, low and high density polyethylene (LDPE, HDPE) used to produce shopping bags.

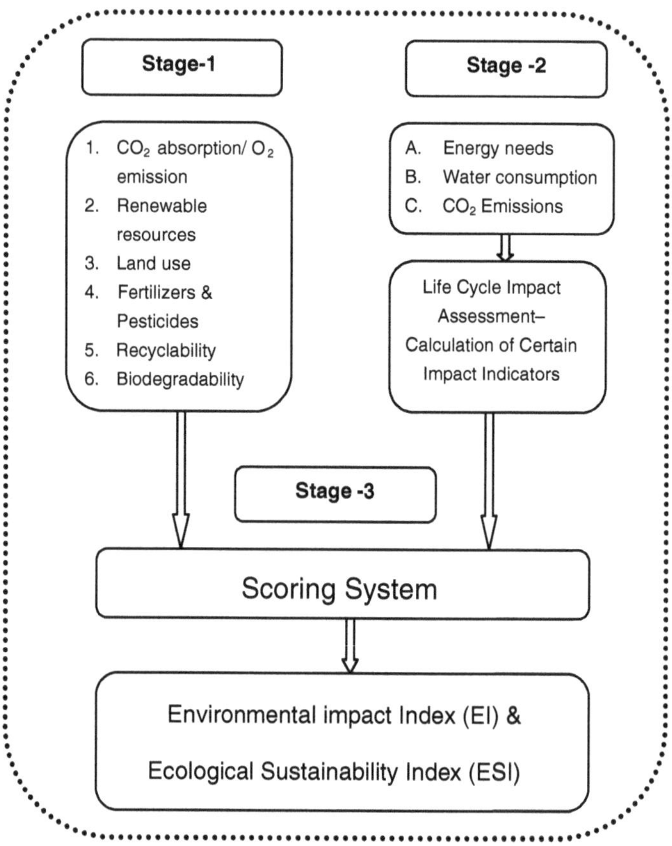

Fig. 3.1 Development of EI, ESI derivation model for raw materials [5]

Significant factors, which govern the environmental impact during the manufacturing stage of raw materials from the extraction stage, were focused to develop this model. This model was developed in three stages, as shown in Fig. 3.1.

In the first stage, the following factors were considered: the amount of oxygen produced/carbon dioxide absorbed consequently contributing to off-set global warming during the production phase of a fibre, utilization of renewable resources, land use, usage of fertilizers and pesticides, fibre recyclability and biodegradability of chosen fibres and other raw materials employed to produce grocery shopping bags.

In the second stage, a different consideration to environmental impact was given with the aid of life cycle assessment. With the consideration of very important factors namely the amount of energy consumed, quantity of water utilized and amount of green house gases emitted as life cycle inventory (LCI), a life cycle impact assessment (LCIA) study, which will elucidate the characteristics of ecological sustainability was conducted to derive certain impact categories pertaining to the damage caused to human health, ecosystem quality and resources.

In the third stage, a scoring system based on the above mentioned factors, which predominantly determine ecological sustainability was framed, from which an Environmental Impact Index (EI) was developed. Further, an Ecological Sustainability Index (ESI) was derived from the EI values for the chosen fibres and other raw materials. The ecological sustainability index can be mathematically expressed as follows [5]:

$$\mathrm{EI} = \sum \alpha_j \mathrm{Y}_j = \alpha_1 \mathrm{Y}_1 + \alpha_2 \mathrm{Y}_2 + \alpha_3 \mathrm{Y}_3 + \alpha_4 \mathrm{Y}_4 + \alpha_5 \mathrm{Y}_5 + \alpha_6 \mathrm{Y}_6 + \alpha_7 \mathrm{Y}_7. \quad (3.1)$$

$$\mathrm{ESI}_k = (1 - \mathrm{EI}_k/\mathrm{EI}_{max}) \times 100. \quad (3.2)$$

where,

EI	Environmental Impact index
EI_k	Environmental impact index of the *k*th fibre under consideration
EI_{max}	The gained maximum scores of Environmental impact index among the selected fibres
ESI	Ecological Sustainability Index
ESI_k	Ecological Sustainability Index of the *k*th fibre under consideration
α_j	Weighting coefficient for the *j*th factor
Y_1	CO_2 absorption/ O_2 emission in fibre production ready for textile processing
Y_2	Use of renewable resources in fibre production
Y_3	Land use in fibre production ready for textile processing
Y_4	Usage of fertilizers and pesticides in fibre production
Y_5	Fibre recyclability
Y_6	Fibre biodegradability
Y_7	EI_{LCIA}–LCIA Impact categories, which is defined as $\mathrm{Y}_7 = \sum \beta_i X_i = \beta_1 X_1 + \beta_2 X_2 + \beta_3 X_3 (X_1, \ldots X_3) = \mathrm{f}(\mathrm{x}_1, \mathrm{x}_2, \mathrm{x}_3)$, i.e. $X_1 = \mathrm{f}_1(\mathrm{x}_1, \mathrm{x}_2, \mathrm{x}_3)$
β_i	Weighting coefficient for the *i*th LCIA indices
X_1	Damage to Human Health
X_2	Damage to Eco System Quality
X_3	Damage to Resources
x_1	Energy consumption in fibre production ready for textile processing
x_2	Water consumption in fibre production ready for textile processing
x_3	CO_2 Emissions in fibre production ready for textile processing.

According to the developed model, organic cotton was the most preferred raw material and acrylic the least. Other raw materials for grocery bags such as LDPE, HDPE and paper lie between organic cotton and acrylic. A sensitivity study was also conducted to check the robustness of the developed model.

3.3.1 Details of Stage 1

3.3.1.1 Amount of Oxygen Produced/CO_2 Emitted

One of the principal factors that decides the ecological sustainability is the quantity of oxygen released to the atmosphere, hence it is considered in this model with topmost priority. It is a renowned fact that there exists a process which converts CO_2 into organic compounds such as sugar, by consuming energy from sunlight in the presence of water, during this process besides sugar; oxygen is released as a byproduct and it is pivotal for life on earth for all living organisms [6].

This is a priceless gift to humanity from plants (so are the fibres and other materials derived from these plants). This effect primarily needs to be taken into account for determining ecological sustainability. CO_2 is eventually absorbed during the whole process and this consequently reduces global warming. These effects are calculated for textile fibres and other raw materials made out of such resources. According to literature, cotton releases 8,000 Kgs of Oxygen / Hectare [7] and absorbs 23,404 Kgs of CO_2 /acre [7]. Hemp absorbs 5,319 Kgs of CO_2 / acre [8]. Viscose releases 2,800 Kgs of O_2/acre/year [8]. Viscose and the paper raw material be used for shopping bag production absorbs 1,000 Kgs of CO_2 /acre [9]. The whole effect of photosynthesis is quite applicable to fibres and raw materials extracted from natural resources such as plants and trees. All other fibres of both animal and synthetic origins do not come into picture and they only emit green house gases such as CO_2 and methane (sheep [10]), which largely contribute to global warming. Keeping these factors in mind, the scoring system pertaining to this category is shown in Fig. 3.2 and the relevant score of each fibre for this category (Y_1) is listed in Table 3.1.

3.3.1.2 Renewable Resources Utilisation

This second category examines the utilisation of resources, whether they are renewable or non-renewable for the production of fibres and other raw materials. Renewable resources are the ones which are replaced by natural processes at a rate comparable or faster than their rate of consumption by humans [11]. With this background, textile fibres obtained from natural resources say plants, trees and animals are renewable (cotton [12], viscose [13], hemp [12], wool [12] and paper) and on the other hand fibres from petroleum sources and other resources (nylon, polyester, polypropylene, LDPE, HDPE and acrylic) [12] which cannot be renewed again are non-renewable. The scoring scheme for this category and the corresponding value of each raw material is given in Fig. 3.2 and Table 3.1 respectively.

Fig. 3.2 Scoring system

CO_2 absorption /emission

Amount of CO_2 absorbed / Hectare/year	Score
<1000	-1
1000-5000	-2
5000-10000	-3
10000-20000	-4
>20000	-5
Negative contribution — CO_2 emission	5

Resources Utilization

Resources	Score
Renewable	-5
Non-renewable	5

Land Usage

Usage of Land	Score
Direct	5
Indirect	1

Usage of synthetic fertilizers and pesticides

Usage	Score
Yes	5
No	1

Recyclability

Recyclability	Score
With Ease	1
With Difficulty	5

Biodegradability

Biodegradability	Score
Yes	1
No	5

Table 3.1 Values for Y_1–Y_6

Fibre	Value of Y_1	Value of Y_2	Value of Y_3	Value of Y_4	Value of Y_5	Value of Y_6
Cotton	−5	−5	5	5	5	1
Organic Cotton	−5	−5	5	1	5	1
Wool	5	−5	5	5	1	1
Hemp	−3	−5	5	5	5	1
Nylon 6	5	5	1	1	1	5
Nylon 66	5	5	1	1	1	5
Polyester	5	5	1	1	1	5
PP	5	5	1	1	5	5
Acrylic	5	5	1	1	5	5
Viscose	−2	−5	5	1	5	1
Paper	−2	0	5	0	5	0
LDPE	5	5	2.5	0	5	5
HDPE	5	5	2.5	0	5	5

3.3.1.3 Land Use

Usage of land as a resource for fibre or raw material growth is the third factor considered. Land use is applicable for natural fibres of cellulosic and animal origin, since they need land for their growth and as well for further processing of fibres into a useful textile product. As a matter of fact, even synthetic fibres need land for their production, for instance the infrastructure of a production base including industrial plants is an indirect usage of land. This category considers both direct and indirect usage of land for growth/ production of fibres. With this background, the scoring scheme and the score of each raw material is given in Fig. 3.2 and Table 3.1.

3.3.1.4 Usage of Synthetic Fertilizers and Pesticides

This category considers the usage of fertilizers and pesticides for the growth of fibres and takes into account for the calculation of EI and ESI. This factor is relevant for natural fibres only, again fibres of plant origin and animal origin (Sheep for instance, needs pesticides in feed or on the pasture land). Scoring scheme and the relevant value of each raw material is given in Fig. 3.2 and Table 3.1.

3.3.1.5 Fibre Recyclability and Biodegradability

The last set of factors represents the behaviour of raw materials in the end-of-life stage. The first one is fibre recyclability, which refers to the ability of the fibre to be recycled and converted to a useful product. Recycling refers to the conversion of the old ones into new products which is discarded after use or otherwise will go to landfill. This aids in the reduction of wastage of materials which has the potential to be used again and to reduce the consumption of fresh raw materials and results in other associated benefits like reduced cost, energy, pollution, etc. Fibres such as cotton, paper and viscose are difficult to recycle and fibres like wool, nylon and polyester are easy to recycle [12]. PP, LDPE, HDPE and acrylic are also difficult to be recycled [14].

Next factor in this category is biodegradability. After use, when textiles are buried in soil, soil-resident microorganisms participate in the degradation of textile materials, which is called biodegradation, and the biodegradability is often used as a standard measurement of the environmental friendliness of textile products [15]. Fibres of natural origin such as cotton, wool, paper and viscose are biodegradable [12]. Nylon and Polyester are non- biodegradable [12]. PP, LDPE, HDPE and acrylic are also belong to non-biodegradable category [14].

Fibre recyclability and biodegradability refer to end-of-life option at the grave stage of products and largely contribute to solid waste management and enable us to dispose/ recycle them in a safe manner. Hence a special consideration is given here in this model for these two factors. Scoring scheme developed and the corresponding value of raw materials under discussion is given in Fig. 3.2 and Table 3.1.

3.3.2 Details of Stage 2

Next stage deals with the quantification of LCIA (life cycle impact assessment). Suitable selected impact categories were earmarked which can reflect the environmental impact and ecological sustainability. As stated earlier, factors such as energy needs, water requirements and CO_2 emissions in fibre production stage were considered as LCI and LCIA was performed with the aid of Simapro 7.2 version of LCA software.

3.3.2.1 Artificial Energy Needs

Confining to the discussion of the artificial energy sources employed in fibre production (natural source of energy from sun light in fibre growth stage was not included), the amount of energy needed to produce one kilogram of fibre is tabulated in Table 3.2 for various fibres.

Table 3.2 Energy, Water needs and CO_2 emission of various raw materials (cradle to gate)

Fibre	Energy use in MJ per kg of fibre	Water requirement Per kg of fibre	CO_2 Emission kg CO_2 Per Kg of Fiber
Nylon 6	120.47 [19]	185 kg [19]	5.5 [19]
Nylon 66	138.65 [19]	663 kg [19]	6.5 [19]
Viscose	100 [18]	640 L [20]	9 [21]
Acrylic	175 [18]	210 L [20]	5 [21]
PET	125 [18]	62 kg [19]	2.8 [19]
Organic Cotton	54 [16]	24,000 kg [16]	2.5 [16]
Wool	63 [18]	125 L; 5–40 L (Scouring) [23]	2.2 [21]
Conventional Cotton	60 [16]	22,000 kg [16]	6 [16]
Hemp	10 [17]	214 L [20]	3.8[22]
PP	115 [18]	43 kg [19]	1.7 [19]
Paper	21.6 [23]	300 L [23]	3.24[23]
LDPE	78.08 [19]	47 kg [19]	1.7 [19]
HDPE	76.71 [19]	32 kg [19]	1.6 [19]

For natural fibres, amount of energy in mega joules required till the production of a particular fibre in mill (field to mill gate) and in case of synthetic fibres, this is the energy utilised from raw material extraction to polymerisation stage until converted to a spinnable fibre.

3.3.2.2 Water Requirements

One more factor which is equally significant as energy needs in determining sustainability is the water quantity needed for the production of raw materials. Table 3.2 enumerates the water requirements to produce one kilogram of different raw materials under question. Here water requirements till the conversion of useful fibre or raw materials are indicated. For fibres of natural—plant and animal origin, the water requirements from initial stage till mill stage and in case of synthetic fibres water needs from raw material extraction to fibre production stage are indicated here. Water requirements listed in Table 3.2 include both processing and cooling needs.

3.3.2.3 CO_2 Emission from Fibres (Cradle to Gate of Fiber)

In this part, the CO_2 emission from fibres in their "cradle to gate stage" only is considered as a factor to determine sustainability and they are tabulated in Table 3.2.

By considering energy needs, water requirements and CO_2 emissions in the production stage of different raw materials considered for this study, life cycle impact assessment was calculated by using SIMAPRO 7.2 version of LCA software. Among the various impact assessment methods available, Eco-indicator'99 (Hierarchist version) method was selected to calculate the damage created by the fibres and other raw materials in the following categories, which can help to evaluate the environmental impact and the sustainability of the fibre production process:

- Damage to Human Health (DALY) (Disability-Adjusted Life Years)
- Damage to Eco System Quality (PDF*m2 yr) (Potentially Disappeared Fraction of plant species)
- Damage to Resources (MJ Surplus) (Additional energy requirement to compensate lower future ore grade).

LCIA was performed using Eco-indicator'99, H/H version method and the results of the same are shown in Table 3.3. The scoring system based on the LCIA impact categories/indicators are depicted in Fig. 3.3 and the corresponding indices of this category are listed in Table 3.3.

3.3.3 *Stage 3*

According to the scoring schemes explained in Figs. 3.2 and 3.3 scores were calculated for the chosen textile fibres and other raw materials used for shopping bags, which are represented by the values of Y_1, Y_2 ....Y_7. According to Eq. (3.1), summation of all the scores in each category results in an index called, "Environmental Impact index (EI)". The EI values of ten fibres and other raw materials

Table 3.3 Life cycle impact assessment results

Fibre	Damage to human health (DALY) (scale:1000:1)		Damage to eco system quality (PDF*m2 yr)		Damage to resources (MJ Surplus)		Value of Y_7
	Result	Score	Result	Score	Result	Score	Score
Cotton	0.5	2	3.2	2	9.4	2	6
Organic Cotton	0.4	2	2.9	2	8.5	2	6
Wool	0.5	2	3.4	2	9.9	2	6
Flax	0.08	0	0.5	0	1.6	0	0
Nylon 6	1	4	6.5	4	18.9	4	12
Nylon 66	1.1	4	7.5	4	21.7	5	13
Polyester	1	4	6.8	4	19.6	4	12
Polypropylene (PP)	0.9	3	6.2	4	18	4	11
Acrylic	1.4	5	9.5	5	27.4	5	15
Viscose	0.8	3	5.4	4	15.7	4	11
Paper	0.17	2	1.2	1	3.4	1	4
LDPE	0.63	3	4.23	3	12.2	3	9
HDPE	0.62	3	4.15	3	12	3	9

Fig. 3.3 Scoring system based on LCIA indicators

Damage to Human Health (DALY)	
<0.1	0
0.11- 0.3	1
0.31-0.6	2
0.61-0.9	3
0.91-1.2	4
>1.21	5
Damage to Eco System Quality (PDF*m2yr)	
<0.5	0
0.6-2	1
2.1-4	2
4.1-6	3
6.1-8	4
>8.1	5
Damage to Resources (MJ Surplus)	
<2	0
2.1-5	1
5.1-10	2
10.1-15	3
15.1-20	4
>20.1	5

under consideration are listed in Table 3.4 by considering all weighing co-efficients as equal. Table 3.4 gives information about Ecological Sustainability Index (ESI) derived for each fibre from EI values using Eq. (3.2). Both EI and ESI values of all ten fibres and other raw materials under discussion are listed in Table 3.4.

The interpretation of EI value goes like this: higher the value, greater the impact the fibres have on the environment. From Table 3.4, it is well evident that organic cotton causes the lowest impact and acrylic causes the highest impact on the environment. As regards to other raw materials for the production of shopping bags such as paper, LDPE and HDPE, paper scores much ahead of its counterparts. Paper stands at an equal position of that of flax and a better next position of organic cotton, because of its merits being a natural, renewable and biodegradable resource, apart from scoring on other considered factors. LDPE and HDPE scored equally and their position lie between PP and nylon 66.

As far as the synthetic fibres like nylon6, 66, polyester, polypropylene, LDPE, HDPE and acrylic are considered, because of their non-biodegradability, utilisation of non-renewable resources and large CO_2 emissions, they could not score as good

Table 3.4 EI and ESI of textile fibres/other raw materials

Fibre	EI	ESI
Cotton	16	57
Organic cotton	11	71
Wool	21	44
Flax	12	68
Nylon 6	29.5	21
Nylon 66	30.5	19
Polyester	29.5	21
PP	33.5	11
Acrylic	37.5	0
Viscose	19	49
Paper	12	68
LDPE	31.5	17
HDPE	31.5	17

as fibres made of biodegradable materials derived from natural renewable resources in terms of EI values. Although at times, they prevail over natural fibres in certain areas considered in this research work, the above mentioned qualities make them earn high EI scores, such as 29.5 by nylon 6 and polyester, 30.5 by nylon 66, 33.5 by polypropylene, 31.5 by LDPE, HDPE and 37.5 by acrylic which implicate a very high environmental impact. Despite their lower amounts of water requirements, their higher energy needs and CO_2 emissions result in higher Y_7 values.

Natural fibres and other raw materials such as conventional cotton, organic cotton, paper, flax, wool score better than the synthetic fibres in terms of ESI values. Among these, organic cotton tops the list among the chosen fibres with an ESI of 71 followed by paper and flax, which gained an equal ESI of 68.

Synthetic fibres like nylon6, 66, polyester, polypropylene, LDPE, HDPE and acrylic scored less ESI values, since they are not as ecologically sustainable as fibres made out of biodegradable materials derived from natural, renewable resources. They obtained low ESI values such as 17 by LDPE, HDPE, 19 by Nylon 66, 21 by nylon 66 and polyester, 11 by polypropylene and 0 by acrylic which implicate low ecological sustainability. In this category, nylon 6 and polyester are found to be better, followed by nylon 66 [24].

3.4 Environmental Impact Assessment of Manufacturing Processes of Shopping Bags

A generalized picture of life cycle assessment and different phases involved in a LCA of a grocery shopping bag is depicted in Fig. 3.4. The previous section dealt with the environmental impact assessment of raw materials. This section deals with the next phase of life cycle of a shopping bag, i.e. manufacturing process.

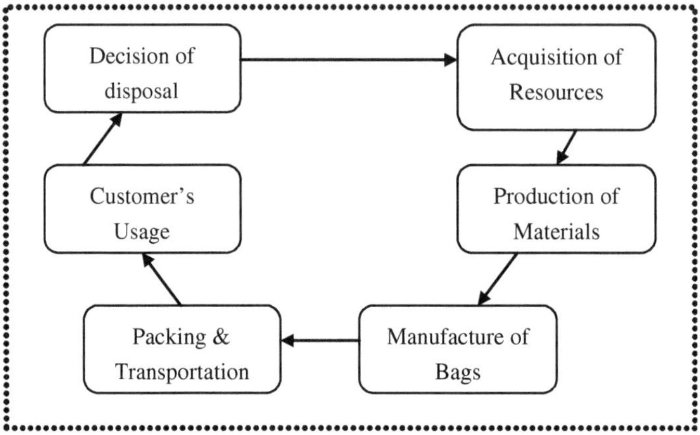

Fig. 3.4 Different life cycle assessment phases of grocery shopping bags

Grocery Shopping bags, being perceived as a symbol of throw-away society, demand LCA to assess their environmental impacts. A large number of studies have been conducted to explore the life cycle impacts created by various shopping bags [25–37]. Most of the studies focused on plastic and paper bags, but very little focus on LCA has been given for nonwoven and woven bags [38].

Life cycle inventory details for various types of grocery bags collected after investigation of various secondary data sources and a further life cycle impact and eco-fucntional assessment of different grocery bags will be dealt with in the forthcoming chapters. This section aims at discussing about a LCA study conducted by collecting primary data in a nonwoven shopping bag manufacturing plant in south china. This LCA study examined the environmental performance of the nonwoven polypropylene shopping bags and its different variants and also highlights the investigated hot-spots in the production process of shopping bags.

A comprehensive life cycle inventory analysis was conducted in a factory manufacturing nonwoven shopping bags for about a week's time. The study was conducted in one of the leading nonwoven manufacturing companies in China, "National Bridge Industrial (Shenzhen) Co., Ltd", situated in Shenzhen, China. This production facility of nonwoven bags has two manufacturing set-ups. One set-up produces nonwoven fabrics (primarily spun bonded) which are transported to another set-up, where the garment manufacturing operations take place, such as spreading, cutting, screen printing, sewing and packaging. This factory has a patented technology, utilizing thermal means to attach the cut pieces of fabrics to produce a shopping bag. This was aimed to replace the conventional sewing technology.

In the production processes of shopping bags, sewing is one of the widely used techniques used to join the separated (cut) parts with stitches to form a complete shopping bag. In case of shopping bags, two sides of a bag are sewn together and also the handles are attached to the body of the bag at sewing stage. The same

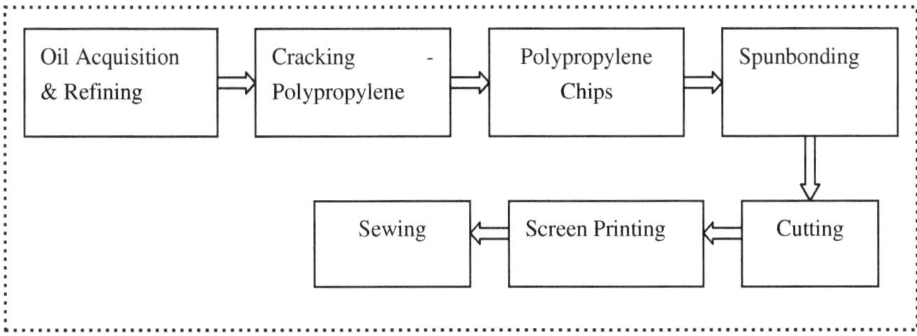

Fig. 3.5 Manufacturing process of nonwoven bags—sewing technology Product A [39]

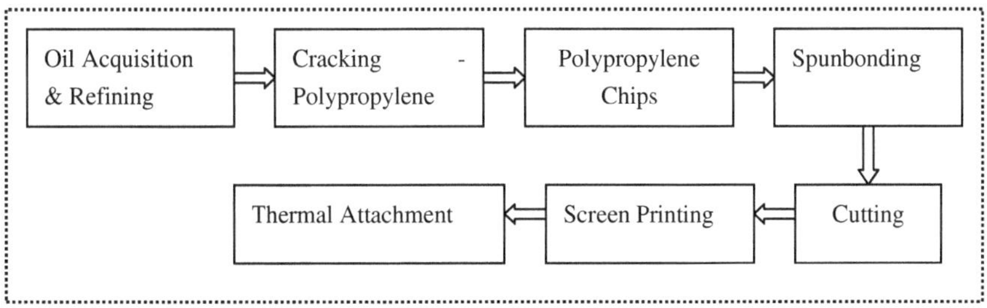

Fig. 3.6 Manufacturing process of nonwoven bags—thermal attachment Product B [39]

operation can be replaced by a thermal technology with the aid of a patented method used by this factory, where very high temperature is used as a means to achieve this operation. Since it is a technology patented by the industry, where the field study took place, details of this technology cannot be disclosed here in accordance with the request of the industrial partners. This research work describes the environmental performance assessment of nonwoven polypropylene shopping bags produced by the two methodologies discussed above.

As stated earlier, a large number of studies have been conducted in the area of environmental performance of shopping bags to investigate their Life Cycle Assessment (LCA). However, as said earlier, very little work has been done on nonwoven and woven bags compared to plastic and paper bags [38]. Even among a very few studies published about nonwoven bags, a comprehensive life cycle inventory is not available for nonwoven bags exclusively and there is a dearth of articles analysing the hot-spots in the production processes of nonwoven bags [39]. Being the first work of this kind, this present study outlined the detailed life cycle inventory of nonwoven shopping bags manufacturing process and also explained the results. The following product types are considered for this LCA study and their processing sequence is explained in Figs. 3.5, 3.6.

1. **Product A—Sewn bag**, (Fabric weight: 100 g/m^2; Size: 43(L)*38(H)*24(D) cm;)
2. **Product B—Thermo bonded with Cutting** (Fabric weight: 75 g/m^2; Size: 36(L)*42.5(H)*19.5(D) cm;)

3.4.1 Life Cycle Assessment of Nonwoven Shopping Bags

This study primarily aimed at collecting the inventory details of cradle to gate stage of production processes of two selected products produced by two different manufacturing techniques. This study also wanted to quantify the environmental impacts created by these two selected products and also to locate the hot-spots which were responsible for major environmental impacts. This study also aimed at making a comparative analysis between sewing and thermal technologies in terms of environmental impacts made.

This study spans from cradle to gate stages including the procurement of PP (Polypropylene) chips, master batch, spun bonding process, transport to cutting process, apparel manufacturing processes such as spreading, cutting, screen printing, sewing/thermal attachment, packaging. Inventory data for the production of PP, master batch and other associated materials were obtained from the eco-invent dataset library from SIMAPRO software. Original transportation details of PP and master batch from the manufacturing plants to this factory were not included in this study. Final transport of shopping bags to customers was also out of scope of this study.

The life cycle inventory (LCI) of both products A and B considered in this study is given in Table 3.5. For both the products, a transportation distance of 6 kms from spun bonding factory and 15 kms for chemicals and other ancillaries for cutting, screen printing process was applicable and considered accordingly into the calculation. Road transport by means of diesel trucks was applicable to both of the products in this study [39].

Environmental Performance assessment (Eco-damage by Eco-indicator'99 method and carbon footprint by IPCC GWP method) was performed with the aid of SIMAPRO 7.2 software. In Eco-Indicator'99, a hierarchist version (V2.06) was used to quantify the life cycle impacts. Carbon footprint assessment was simulated to calculate the global warming potential (GWP) for 100 and 20 years.

The results of the eco-damage assessment by eco-indicator'99 and carbon footprint assessment are presented in Figs. 3.7, 3.8.

Life Cycle Impact Assessment includes different steps such as characterization, normalization, weighing and single score assessment. Figure 3.7 depicts the final step of this assessment, i.e. single score measurement of two types of shopping bags selected for this study. Single score is the measure used to deduce the final result of analysis after comparing different products in a LCA study. Though it cannot be used as a measure to market/display the environmental characteristics of

Table 3.5 Inventory of 1 unit of Products A and B [39]

Inventory details	Product A[a]	Product B[b]
1. *Spun-bonding process*:		
Inputs:		
PP chips	82.12 g	64.2 g
Master batch	1.16 g	0.91 g
Electricity		
Manufacturing	0.0892 kWh	0.0697 kWh
Lighting	0.00163 kWh	0.00127 kWh
Cleaning	0.0002 kWh	0.00015 kWh
Water (cleaning)	1.01 g	0.79 g
NaOH (cleaning)	0.0021 g	0.0016 g
Paper tubes	2.97 g	2.32 g
Plastic sheet (PE)	0.58 g	0.45 g
Outputs:		
Fabrics—standard quality	79.7 g	62.3 g
Fabrics in low quality and multi colour ones	2.37 g	1.85 g
Fabrics-waste	3.6 g	2.82 g
2. *Cutting*		
Inputs:		
Spunbonded fabrics	79.7 g	62.3 g
Electricity	0.00267 kWh	0.00267 kWh
Outputs:		
Cut pieces of fabrics	75.4 g	51.67 g
Waste fabrics	4.26 g	10.58 g
3. *Screen printing*		
Inputs:		
Fabrics (PET mesh) for screen	1.44 grams	0.72 grams
Aluminum for screen	3.34 inches	1.67 inches
Wood for screen	0.0001 inches	0.00005 inches
PE film	0.3 g	0.2 g
Printing ink	3.3 g	3.73 g
Electricity (lighting and fan)	0.0178 kWh	0.0178 kWh
Silicone spray	0.16 g	0.16 g
ABS-Cyanoacrylate	0.06 g	0.06 g
Cyclohexanone	3 g	3 g
Autotype plus 7,000 direct emulsion	0.4 g	0.4 g
Isophorone	0.65 g	0.65 g
Adhesive	2.5 g	2.5 g
Water (cleaning)	0.63 g	0.63 g
Fluid waste (water)	45.8 g	45.8 g
Solid waste (chemicals and others)	4.17 g	4.17 g
4. *Sewing*:		
Electricity	0.0081 kWh	NA
Thread used	0.5 g	NA
5. *Thermal bonding*:		
Electricity	NA	0.0305 kWh

(continued)

Table 3.5 (continued)

Inventory details	Product A[a]	Product B[b]
Fabric waste	NA	NIL
6. *Packaging*:		
Paper box	8.21 g	8.21 g
Plastic sheet (PE)	0.5 g	0.5 g

[a] Weight of 1 bag: 79.2 g
[b] Weight of 1 bag: 55.4 g

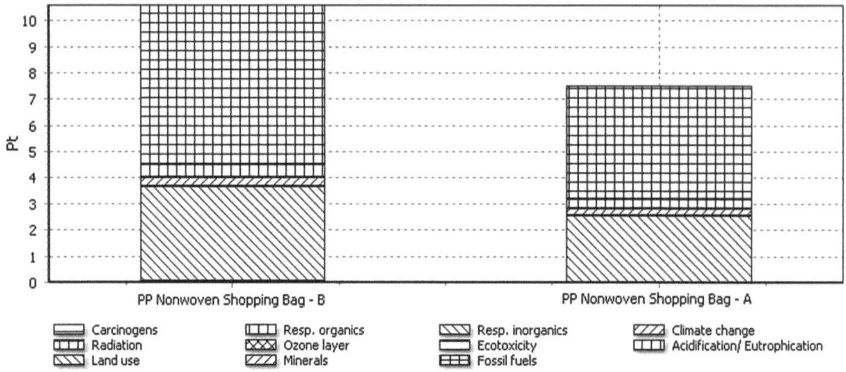

Fig. 3.7 Eco-damage assessment of two products A and B [24]

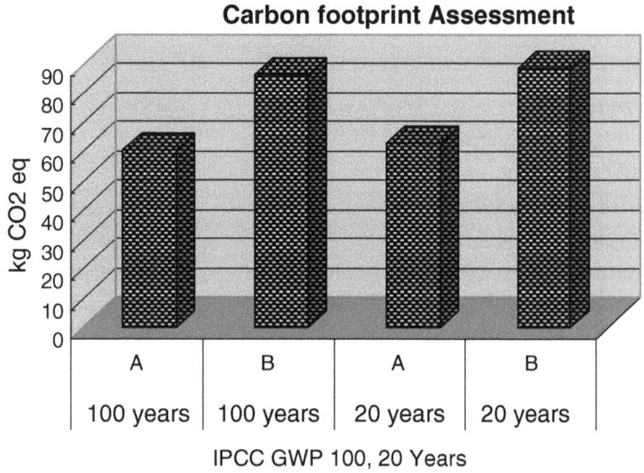

Fig. 3.8 Carbon footprint assessment of two products A and B

comparable products, it certainly can provide an indication of different products' environmental scores.

While comparing products A and B in terms of eco-indicator damage points, the former scores out the later in terms of the environmental damage created, because of the lower level of energy consumed during the manufacturing phase and owing to the lower quantity of waste fabric produced during manufacturing process. Product A surpasses B in terms of various inputs for its comparative unit weight, which is well evident from Table 3.5. It is evident and can be understood from the results that the sewing technology seems to be more efficient and less energy intensive compared to the thermal technology.

When it comes to major hot-spots for eco-indicator assessment, the following elements attributed to the major level of impacts in order of hierarchy (with 1 % cut-off) in eco-damage measurement.

1. Distillate Fuel oil
2. Diesel truck
3. Polypropylene chips manufacturing process
4. Electricity from hard coal
5. Other processes.

The order and elements remain same for both products A and B except for the amount of impact varies for different elements for both of the products.

The life cycle impact assessment results for carbon footprint are shown in Fig. 3.8.

From the results displayed in Fig. 3.8, it can be noticed that product A produced by conventional sewing technology created a significantly lower carbon footprint results than product B, produced by thermal means. For both the products studied, transportation by diesel trucks and the consumption of electricity for the production process of shopping bags as well as the energy intensive PP chips manufacturing process were found to be the major threats to global warming [39].

In this study, a comparative life cycle inventory (cradle to gate) was obtained for two polypropylene nonwoven bags manufactured by two different technologies—conventional sewing technology and thermal attachment from a manufacturing setup located in Southern China. The hot-spots in the manufacturing processes of the two selected products and also a comparative life cycle impact assessment study to assess the eco-damage and carbon footprint created by two types of bags were also performed. Environmental impacts were quantified from the cradle to gate stage of polypropylene nonwoven bags and within certain boundaries indicated earlier.

Out of the two major technologies involved in manufacturing nonwoven shopping bags in the attachment phase assessed in this study, it was clear that sewing technology was better in terms of environmental damage and carbon footprint than thermal technology. Product A, though it assumes more inputs in the spun-bonding process due to its higher unit weight, outscores B due to its lower energy requirements, low level of waste creation and other related factors in terms of comparative unit weight.

With regards to major hot-spots, transportation by diesel truck, manufacturing process impacts of polypropylene and the consumption of electricity are the major elements that impact the environment. As far as the production of polypropylene and the printing colour/dye used, nothing much can be done at the user's end, except advising the supplier from whom these are procured to take care of the environmental issues pertaining to the manufacturing impacts of the respective products. Also it was advised to procure from the closest manufacturers, though transportation impact is not included in this study.

Local transportation was found to be one of the major threats of environmental impacts and this is a two-fold issue. One is transporting the spun bonded fabrics to the cutting factory and the other one is procuring chemicals and other essential items for production. Though it is obligatory to transport spun bonded fabrics from one station to the other, it was strongly recommended to look for alternative renewable energy measures to curtail the negative impacts on the environment. It was also advised to look for a nearest dealer to reduce the transportation impact as far as procurement of chemicals and other items are concerned. In general, it is worthwhile implementing better alternatives/technologies to reduce energy consumption and hence its corresponding environmental impacts [24, 39].

Though the majority of fabric waste is recycled, this study does not include the usage of recycled PP in its manufacturing phase, since it is difficult to account for calculations. If this was included, the impact assessment results may well be different and the impacts will certainly be reduced.

It was recommended to seek for alternative technologies in order to reduce power consumption, to reduce the energy impacts and an energy audit was also recommended for this factory, where the study was conducted.

3.5 Life Cycle Assessment of Shopping Bags: Manual Calculation of LCA Values for Chinese Factors

In life cycle assessment, while quantifying impacts, there are many mandatory and optional elements involved as indicated in ISO 14040 standard. It is an expectation of any reader/assessor/ consumer to understand the impacts of any product pertaining to his or her home country he is from. Environmental impacts pertaining to a particular country is different from others on various grounds and certain impacts are pertinent to only specific countries as well, which will not be applicable to other countries.

In this study, as discussed in previous section, quantification of environmental impacts was performed by one of the well reputed commercial softwares called Simapro, which was produced by PRE Consultants of the Netherlands. Since this simapro software originates from Europe, all the impact values calculated from it are very much relevant to Europe and to people living there. Although SIMAPRO can quantify different impacts worldwide in addition to the specific impacts for Europe, there is no method available till date to explicitly quantify the impacts of consumers in China from any of the LCA softwares available today. This is the case with most of the LCA packages, since many of them originated primarily from Europe. However, it is possible to select the inventory details pertaining to China from different datasets inbuilt with SIMAPRO. However, it is not possible to quantify the characterization and normalization impact values for China alone. Hence an attempt has been made to perform the LCA calculations without the aid of LCA software [24] and the details of this attempt are discussed here. This attempt involved utilization of the well known equations used for the characterization step in LCA to calculate manually the impact values in this section. The characterization and normalization values used in this section are solely applicable to China. Inventory details directly related to China are referred from the latest data source and the details are discussed below.

3.5.1 Life Cycle Inventory Details

This study revolves around the life cycle impact assessment of five different types of shopping bags primarily used for grocery purposes. They were chosen to cover the major categories of shopping bags such as plastic, paper, nonwoven and woven shopping bags and it was a cradle to gate stage study. LCI data for this study were obtained from the secondary data sources and Table 3.6 lists the LCI data for the production of one unit of shopping bag. As it is evident from Table 3.6, major areas covered in LCI are the primary energy used to produce shopping bags and the GHG emitted during the production phase of shopping bags. By considering these two very important factors, the environmental impacts pertaining to the impact values related to China were quantified.

The first and foremost requirement for the calculation of life cycle impacts is the electricity input. To obtain this, the electricity inventory for China is referred to quantify the impacts corresponding to generation of energy in China. The

Table 3.6 LCI of various shopping bags for one unit of bag [27, 28]

Alternative	Weight/ bag (g)	Material consumption (g)	Green house gas emissions (CO_2 eq.) (g)	Primary energy
Plastic bag (HDPE)	6.0	6.0	11.6	0.40 MJ (0.11 kWh)
Paper bag (kraft paper)	42.6	42.6	22.7	1.40 MJ (0.39 kWh)
Boutique plastic (LDPE)	18.1	18.1	45.8	1.47 MJ (0.41 kWh)
Woven cotton bag	125.4	125.4	277.0	17.58 MJ (4.88 kWh)
PP fibre nonwoven bag	65.6	65.6	472.0	11.15 MJ (3.1 kWh)

Table 3.7 Electricity inventory for China

Emission inventory for electricity in China	kg/kWh	g/kWh
Consumption of coal	0.457	457
Consumption of oil	0.0088	8.8
Consumption of gas	0.00795	7.95
Consumption of enriched uranium	0.0000000903	0.0000903
CO_2	0.877	877
SO_2	0.00804	8.04
NO_x	0.00523	5.23
CO	0.00125	1.25
CH_4	0.00265	2.65
Nonmethane volatile organic compound (NMVOC)	0.000395	0.395
Dust	0.0163	16.3
As	0.00000162	0.00162
Cd	0.0000000103	0.0000103
Cr	0.000000137	0.000137
Hg	0.0000000711	0.0000711
Ni	0.000000203	0.000203
Pb	0.00000142	0.00142
V	0.00000233	0.00233
Zn	0.00000194	0.00194
Emissions of waste water	1.31	1310
COD	0.0000602	0.0602
Coal fly ash	0.0834	83.4
Slag	0.0187	18.7
Halogen	37.4 Bq	37400 Bq
Gasoloid	0.161 Bq	161 Bq
Tritium	42.2 Bq	42200 Bq
Non-tritium	0.0406 Bq	40.6 Bq
Radioactive solid waste	0.000000000268 m^3	0.000000268 m^3

corresponding data pertaining to the electricity generation were taken from the latest possible source [40]. The electricity inventory, i.e. life cycle inventory for electricity generation in China to produce 1 kWh of energy is listed in Table 3.7.

The emission inventory values were calculated from the electricity inventory from the values listed in Table 3.8 for different types of shopping bags in terms of the energy requirement values listed in Table 3.6. The second important input to be considered for this LCA calculation in Table 3.6 is GHG emissions. Merging these two inputs discussed above, Table 3.9 lists the quantified total inventory for the production processes of shopping bags. The results are expressed in kilograms per unit of bag.

Table 3.8 Emission inventory for shopping bags for electricity input per bag

Emission Inventory for electricity in China	g/kWh	Plastic bag (HDPE) 0.11 kWh	Paper bag (kraft) 0.39 kWh	PP fibre non-woven bag 3.1 kWh	Woven cotton bag 4.88 kWh	Boutique plastic (LDPE) 0.41 kWh
Consumption of coal	457	50.27	178.23	1,416.7	2,230.16	187.37
Consumption of oil	8.8	0.968	3.432	27.28	42.944	3.608
Consumption of gas	7.95	0.8745	3.1005	24.645	38.796	3.2595
Consumption of enriched uranium	0.0000903	0.00000993	0.0000352	0.00028	0.000441	0.000037
CO_2	877	96.47	342.03	2718.7	4279.76	359.57
SO_2	8.04	0.8844	3.1356	24.924	39.2352	3.2964
NO_x	5.23	0.5753	2.0397	16.213	25.5224	2.1443
CO	1.25	0.1375	0.4875	3.875	6.1	0.5125
CH_4	2.65	0.2915	1.0335	8.215	12.932	1.0865
NMVOC	0.395	0.04345	0.15405	1.2245	1.9276	0.16195
Dust	16.3	1.793	6.357	50.53	79.544	6.683
As	0.00162	0.000178	0.000632	0.005022	0.007906	0.000664
Cd	0.0000103	0.00000113	0.00000402	0.0000319	0.0000503	0.00000422
Cr	0.000137	0.0000151	0.0000534	0.000425	0.000669	0.0000562
Hg	0.0000711	0.00000782	0.0000277	0.00022	0.000347	0.0000292
Ni	0.000203	0.0000223	0.0000792	0.000629	0.000991	0.0000832
Pb	0.00142	0.000156	0.000554	0.004402	0.00693	0.000582
V	0.00233	0.000256	0.000909	0.007223	0.01137	0.000955
Zn	0.00194	0.000213	0.000757	0.006014	0.009467	0.000795
Emissions of waste water	1310	144.1	510.9	4061	6392.8	537.1
COD	0.0602	0.006622	0.023478	0.18662	0.293776	0.024682

(continued)

Table 3.8 (continued)

Emission Inventory for electricity in China	g/kWh	Plastic bag (HDPE) 0.11 kWh	Paper bag (kraft) 0.39 kWh	PP fibre non-woven bag 3.1 kWh	Woven cotton bag 4.88 kWh	Boutique plastic (LDPE) 0.41 kWh
Coal fly ash	83.4	9.174	32.526	258.54	406.992	34.194
Slag	18.7	2.057	7.293	57.97	91.256	7.667
Halogen	37,400 Bq	4,114 Bq	14,586 Bq	115,940 Bq	182,512 Bq	15,334 Bq
Gasoloid	161 Bq	17.71 Bq	62.79 Bq	499.1 Bq	785.68 Bq	66.01 Bq
Tritium	42,200 Bq	4,642 Bq	16,458 Bq	130,820 Bq	205,936 Bq	17,302 Bq
Non-tritium	40.6 Bq	4.466 Bq	15.834 Bq	125.86 Bq	198.128 Bq	16.646 Bq
Radioactive solid waste	0.000000268 m^3	0.0000000295 m^3	0.000000105 m^3	0.000000831 m^3	0.00000131 m^3	0.00000011 m^3

Table 3.9 Total inventory for production processes of shopping bags per bag in kilograms

Inventory	Plastic bag (HDPE)	Paper Bag (kraft)	PP fibre non-woven bag	Woven cotton bag	Boutique plastic (LDPE)
Consumption of coal	0.05027	0.17823	1.4167	2.23016	0.18737
Consumption of oil	0.000968	0.003432	0.02728	0.042944	0.003608
Consumption of gas	0.0008745	0.0031005	0.024645	0.03879	0.032595
Consumption of enriched uranium	9.93E−09	3.52E−08	2.8E−07	4.41E−07	3.7E−08
CO_2	0.10807	0.36473	3.1907	4.55676	0.40537
SO_2	0.0008844	0.0031356	0.024924	0.0392352	0.0032964
NO_x	0.0005753	0.0020397	0.016213	0.0255224	0.0021443
CO	0.0001375	0.0004875	0.003875	0.0061	0.0005125
CH_4	0.0002915	0.0010335	0.008215	0.012932	0.0010865
Nonmethane volatile organic compound (NMVOC)	0.00004345	0.00015405	0.0012245	0.0019276	0.00016195
Dust	0.001793	0.006357	0.05053	0.079544	0.006683
As	1.78E−07	6.32E−07	5.022E−06	7.906E−06	6.64E−07
Cd	1.13E−09	4.02E−09	3.19E−08	5.03E−08	4.22E−09
Cr	1.51E−08	5.34E−08	4.25E−07	6.69E−07	5.62E−08
Hg	7.82E−09	2.77E−08	2.2E−07	3.47E−07	2.92E−08
Ni	2.23E−08	7.92E−08	6.29E−07	9.91E−07	8.32E−08
Pb	1.56E−07	5.54E−07	4.402E−06	0.00000693	5.82E−07
V	2.56E−07	9.09E−07	7.223E−06	0.00001137	9.55E−07
Zn	2.13E−07	7.57E−07	6.014E−06	9.467E−06	7.95E−07
Emissions of waste water	0.1441	0.5109	4.061	6.3928	0.5371
COD	6.622E−06	2.3478E−05	0.0001866	0.00029378	2.4682E−05
Coal fly ash	0.009174	0.032526	0.25854	0.406992	0.034194
Slag	0.002057	0.007293	0.05797	0.091256	0.007667
Halogen	4.114 Bq	14.586 Bq	115.940 Bq	182.512 Bq	15.334 Bq
Gasoloid	0.01771 Bq	0.06279 Bq	0.4991 Bq	0.78568 Bq	0.06601 Bq
Tritium	4.642 Bq	16.458 Bq	130.820 Bq	205.936 Bq	17.302 Bq
Non-tritium	0.004466 Bq	0.015834 Bq	0.12586 Bq	0.198128 Bq	0.016646 Bq
Radioactive solid waste	2.95E−11 m^3	1.05E−10 m^3	8.31E−10 m^3	1.31E−09 m^3	1.1E−10 m^3

3.5.2 Life Cycle Impact Assessment

3.5.2.1 Characterisation

The assessment begins with the characterization and various environmental categories to be characterized.

Environmental Burden

Environmental burden is being expressed in terms of environmental load units (ELU) and depletion of abiotic resources. ELU is expressed separately for natural resources, emissions to air, fresh water and sea water. The equation for the calculation of environmental burden is as follows [41]:

$$\text{Environmental Burden} = \sum_{i} \text{Factor}_i * m_i \tag{3.3}$$

The total environmental burden is expressed in Environmental Load Units. Factor $_i$ (ELU.kg^{-1}) is the valuation weighting factor for the EPS method for resource i, while m_i (kg) is the quantity of resource i used. The values for these impact categories are tabulated in Table 3.10. All the values listed in Table 3.10 are taken from the handbook of life cycle assessment [41], except the value of gas for ELU-natural resources category, which is taken from another source of Ref. [42]. Table 3.11 lists the Chinese characterization and normalization factors for ADP from the latest Ref. [43]. The equation for the calculation of ADP is as follows [41]:

$$\text{Abiotic Depletion} = \sum_{i} \text{ADP}_i * m_i \tag{3.4}$$

The indicator result is expressed in kg of the reference resource antimony. ADPi is the Abiotic Depletion Potential of resource i, while m_i (kg, except for natural gas and fossil energy) is the quantity of resource i used.

The values for carbon footprint (GWP_{100}), ozone depletion potential, human toxicity for air, fresh water, sea water, agricultural soil and industrial soil are listed in Table 3.12. Values for acidification, eutrophication, radiation and photo- oxidant chemical potential are listed in Table 3.13. The equations for the calculation of carbon footprint (GWP_{100}), ozone depletion potential, human toxicity for air, fresh water, sea water, agricultural soil and industrial soil are as follows [41]:

Carbon Footprint

$$\text{Climate Change} = \sum_{\text{i}} \text{GWP}_{\text{a,i}} * m_i \tag{3.5}$$

The indicator result is expressed in kg of the reference substance, CO_2. GWP $_{a,\, i}$ is the Global Warming Potential for substance i integrated over a specified number of years, while m $_i$ (kg) is the quantity of substance i emitted.

Ozone Depletion Potential

$$\text{Ozone Depletion} = \sum_{\text{i}} \text{ODP}_i * m_i \tag{3.6}$$

Table 3.10 Environmental Burden—ELU and depletion of abiotic resources

Inventory	ELU for natural resources (ELU/kg)	ELU for emissions—air (ELU/kg)	ELU for emissions—fresh water (ELU/kg)	ELU for emissions—sea water (ELU/kg)	ADP (in kg antimony eq./kg)
Coal	0.05				0.0134 (hard coal)
Oil	0.5				0.0201
Gas	1.1 [100]				0.0187 (natural gas) (kg antimony/m^3 natural gas)
Uranium	1,260				0.00287
CO_2		0.0636			
SO_2		0.0545			
NO_x		0.395			
CO		0.191			
CH_4		1.56			
NMVOC					
Dust		0.0071 (PM_{10})			
As	1,900	10			0.00917
Cd	23,000	21.2			0.33
Cr	33	0.8			0.000858
Hg	40,000	177			0.495
Ni	40				0.000108
Pb	240	291			0.0135
V	28.3				1.16E−6
Zn	49				0.000992
Waste water					
COD			0.006		
Coal fly ash					
Slag					
Halogen fluorine (F), chlorine (Cl), bromine (Br), iodine (I), and astatine (At)					
Gasoloid					
Tritium					
Non-tritium					
Radioactive solid waste					

Table 3.11 Chinese characterisation and normalization factors for ADP

Inventory	ADP (in kg antimony eq./kg)	Reserve kg	Normalisation Figures
Coal	7.97×10^{-8}	5.86×10^{7}	0.00214114
Oil	9.91×10^{-5}	8.74×10^{8}	1.15E−07
Gas	8.89×10^{-8}	3.85×10^{8}	0.00029217
Cr	6.31×10^{-2}	6.17×10^{8}	2.57E−10
Hg	7.46	5.97×10^{8}	2.25E−11
Ni	5.65×10^{-2}	4.21×10^{8}	4.20E−10
Pb	4.36×10^{-2}	1.64×10^{9}	1.40E−10
Zn	1.16×10^{-2}	1.12×10^{9}	7.70E−10
Total			0.002433423 4.67×10^{-11}

The indicator result is expressed in kg of the reference substance,CFC-11. ODP $_i$ is the Ozone Depletion Potential for substance i, while m $_i$ (kg) is the quantity of substance i emitted.

Human Toxicity Potential

$$\text{Human toxicity} = \sum_{i}\sum_{ecom} HTP_{ecom,i} * m_{ecom,i} \tag{3.7}$$

The indicator result is expressed in kg 1, 4-dichlorobenzene equivalent. HTP $_{ecom,\ i}$ is the Human Toxicity Potential (the characterisation factor) for substance i emitted to the emission compartment ecom (= air, fresh water, sea water, agricultural soil or industrial soil), while m $_{ecom,\ i}$ is the emission of substance i to medium ecom.

The equations for the calculation of Acidification, Eutrophication, Radiation and POCP are as follows [41]:

Acidification

$$\text{Acidification} = \sum_{i} AP_i * m_i \tag{3.8}$$

The indicator result is expressed in kg SO_2 equivalents. AP $_i$ is the Acidification Potential for substance i emitted to the air, while m $_i$ is the emission of substance i to the air.

Eutrophication

$$\text{Eutrophication} = \sum_{i} P_i * m_i \tag{3.9}$$

Table 3.12 Characterisation values for GWP, ODP and Human Toxicity

Inventory	Climate Change (GWP_{100}) (Air)	Ozone depletion (ODP)	HTP—100 yr (kg 1,4-DCB eq./kg) air	HTP—100 yr (Kg 1,4-DCB eq./kg) fresh water	HTP—100 yr (Kg 1,4-DCB eq./kg) sea water	HTP—100 yr (Kg 1,4-DCB eq./kg) agri soil	HTP—100 yr (Kg 1,4-DCB eq./kg) industrialsoil
Coal							
Oil							
Gas							
Uranium							
CO_2	1						
SO_2			9.6E−02	X	X	X	X
NO_x			1.2E+00	X	X	X	X
CO							
CH_4	21						
NMVOC Benzene/1,1,1-trichloroethane	/110	/0.11	1.9E+03/1.6E+01	1.8E+03/1.6E+01	2.1E+02/9.6E+00	1.5 E+04/1.6 E+01	1.6E+03/1.6E+01
Dust			8.2E−01	X	X	X	X
As			3.5E+05	1.3E+02	3.1E+01	3.1E+02	4.8E+00
Cd			1.5E+05	1.1E+01	6.9E+00	2.8E+03	8.7E+00
Cr							
Hg—Mercury			2.6E+02	1.0E+02	1.2E+02	1.3E+02	9.5E+00
Ni			3.5E+04	4.3E+01	7.8E+00	1.7E+02	3.0E+00
Pb—Lead			2.9E+01	5.2E+00	7.1E+00	2.7E+01	2.4E+00
V			2.6E+02	2.7E+02	4.6E+01	1.3E+03	1.4E+01
Zn			9.6E+01	2.0E−01	2.0E−01	4.5E+00	1.5E−02
Waste water							
COD							
Coal fly ash							
Slag							

(continued)

Table 3.12 (continued)

Inventory	Climate Change (GWP_{100}) (Air)	Ozone depletion (ODP)	HTP—100 yr (kg 1,4-DCB eq./kg) air	HTP—100 yr (Kg 1,4-DCB eq./kg) fresh water	HTP—100 yr (Kg 1,4-DCB eq./kg) sea water	HTP—100 yr (Kg 1,4-DCB eq./kg) agri soil	HTP—100 yr (Kg 1,4-DCB eq./kg) industrialsoil
Halogen fluorine (F), chlorine (Cl), bromine (Br), iodine (I), and astatine (At)							
Gasoloid							
Tritium							
Non-tritium							
Radioactive solid waste							

Table 3.13 Characterisation values for acidification, eutrophication, radiation and POCP

Inventory	Acidification potential (in kg SO_2 -eq./kg)	Eutrophication potential (in kg PO_4^{3-} eq./kg)	Ionising radiation—damage factor ($Yr.KBq^{-1}$) in Air	Ionising radiation—damage factor ($Yr.KBq^{-1}$) in sea water	POCP (in kg ethylene eq./kg)
Coal					
Oil					
Gas					
Uranium					
CO_2					
SO_2	1				0.048
NO_x	0.7	0.13			0.028
CO					0.027
CH_4					0.006
NMVOC Benzene/ 1,1,1-trichloroethane					0.218/0.009
Dust					
As					
Cd					
Cr					
Hg—Mercury					
Ni					
Pb—Lead					
V					
Zn					
Waste water					
COD		0.022			
Coal fly ash					
Slag					
Halogen fluorine (F), chlorine (Cl), bromine (Br), iodine (I), and astatine (At)					
Gasoloid					
Tritium			1.40E−11	6.9E−14	
Non-tritium					
Radioactive solid waste					

The indicator result is expressed in kg PO_4^{3-} equivalent. EP_i is the Eutrophication Potential for substance i emitted to air, water or soil, while m_i is the emission of substance i to air, water or soil.

Photo-Oxidant Formation

$$\text{Photo} - \text{oxidant formation} = \sum_i \text{POCP}_i * m_i \tag{3.10}$$

The indicator result is expressed in kg of the reference substance, ethylene. $POCP_i$ is the Photochemical Ozone Creation Potential for substance i, while m_i (kg) is the quantity of substance i emitted.

Ionising Radiation

$$\text{Radiation} = \sum_{ecomp} \sum_i \text{Damage Factor}_{ecomp,i} * a_{ecomp,i} \tag{3.11}$$

The indicator result is expressed in yr. Damage Factor $_{ecomp,\ i}$ ($yr.kBq^{-1}$) is the characterisation factor substance i emitted to ecomp based on DALYs, while a $_{ecomp,\ i}$ (kBq) is the activity of substance i emitted to compartment ecomp.

The characterization results were calculated according to the values given in Tables listed above step by step and according to the well known life cycle characterization equations discussed above. The results of the characterization step are summarized in Table 3.14.

3.5.2.2 Normalization

The next step in life cycle impact assessment is normalization. Unlike characterization, it is not a mandatory step, but it is very important step, since it normalizes the characterized impact results to an average individual, making the impact assessment results more meaningful. Normalization is done by the following equation [44]:

$$NP(j) = EP(j)/(T \times ER(j)) \tag{3.12}$$

where

EP(j) is the environmental impact potential for impact category j,
NP(j) is the normalised environmental impact potential for impact category j,
T is the expected lifetime of the product in years,
ER(j) is the normalisation reference for impact category j.

Table 3.14 Results of characterization by manual calculation of LCA

Impact category	Plastic bag (HDPE)	Paper bag (kraft paper)	PP fibre nonwoven bag	Woven cotton bag	Boutique plastic (LDPE)
GWP 100 years	1.19E−01	4.03E−01	3.50E+00	5.04E+00	4.46E−01
Ozone depletion potential	4.78E−06	1.69E−05	1.35E−04	2.12E−04	1.78E−05
Human toxicity	6.86E−02	2.45E−01	1.96E+00	3.08E+00	2.59E−01
Acidification	1.28E−03	7.27E−03	3.63E−02	5.71E−02	4.80E−03
Eutrophication	7.49E−05	2.66E−04	2.11E−03	3.32E−03	2.79E−04
Photo-oxidant chemical formation	6.44E−05	2.28E−04	1.82E−03	2.86E−03	2.40E−04
Ionising radiation	6.55E−11	2.32E−10	1.84E−09	2.90E−09	2.44E−10
Radioactive solid waste in m^3	2.95E−11	1.05E−10	8.31E−10	1.31E−09	1.10E−10
Coal fly ash	9.17E−03	3.25E−02	2.59E−01	4.07E−01	3.42E−02
Slag	2.06E−03	7.29E−03	5.80E−02	9.13E−02	7.67E−03
Emissions of waste water	1.44E−01	5.11E−01	4.06E+00	6.39E+00	5.37E−01
Depletion of abiotic resources	7.10E−04	2.52E−03	2.00E−02	3.15E−02	3.19E−03
Environmental burden-ELU-emissions	7.69E−03	2.61E−02	2.26E−01	3.26E−01	2.88E−02
Environmental burden-ELU-resources	4.71E−03	1.67E−02	1.33E−01	2.09E−01	4.98E−02
ADP-Chinese factors	1.17E−06	4.16E−06	3.24E−05	5.21E−05	4.40E−06

Thus, as a result of normalisation, all environmental impacts from the product are expressed as a fraction of an average person's yearly contribution to the impact, and the unit is milliperson equivalents, mPE.

$$\text{So, NP(j)} = \text{EP(j)}/(1 \times \text{ER (j)}) = \text{EP(j)}/\text{ER(j)} \tag{3.13}$$

Table 3.15 lists the values for normalization and weighting pertaining to China [44].

Chinese Factors for ADP are calculated by the following equation [43]:

$$N = 1/\sum_{2004} R_i * ADP_i \tag{3.14}$$

where N is normalization factor of abiotic resource depletion, Ri is reserves of the resource i (kg), $ADPi$ is characterization factors for the resource i (kg antimony eq./kg), 2004 is the benchmark time the year of 2004.

On the basis of relative reserves of China's major resources, it can be calculated that the total resource reserves in 2004 are equal to 2.14 × 1010 kg antimony eq.and the normalization factor for resource depletion is therefore 4.67 × 10−11 [43].

Normalized Results

Tables 3.5, 3.6, 3.7, 3.8, 3.9, 3.10 below lists the normalized results for several impact categories from the values taken from Table 3.15. Tables 3.5, 3.6, 3.7, 3.8, 3.9, 3.10, 3.11 shows the unnormalized results, for the impact categories where normalization values for China are not currently available (Tables 3.16, 3.17).

Weighting

The next step in impact assessment is weighting, which is done by the following equation [44]:

$$\mathrm{WP}\ (j) = \mathrm{WF}\ (j) \times \mathrm{NP}\ (j) \tag{3.15}$$

where

WP (j) is the weighted environmental impact potential for impact category j and
WF (j) is a weighting factor for environmental impact category j

The results of weighting are shown in Table 3.18.

Table 3.15 Normalization and weighting values for China [44]

Impact category	Normalisation reference, ER90[c]				Normalisation reference unit	Weighting factor WFT2000[d]
	East	Central	West	China in total		
Global warming[a]	8,700[e]				kg CO_2 eq/person/year	0.83
Ozone depletion[c]	0.20[e]				kg CFC11 eq/person/year	2.7
Acidification[b]	35	33	41	36	kg SO_2-eq/person/year	0.73
Nutrient enrichment[b]	57	60	67	61	kg NO_3-eq/person/year	0.73
Photochemical ozone formation[b]	0.76	0.63	0.48	0.65	kg C_2H_4-eq/person/year	0.4
Bulk waste[b]	291	247	186	251	kg bulk waste/person/year	0.62
Hazardous waste[b]	22	17	15	18	kg hazard. waste/person/year	0.45
Slag and ashes[b]	18	21	16	18	kg slag and ashes/person/year	0.61

[a] Reference region: World. [b] Reference region: East China, Central China, West China or China in total. [c] Reference year: 1990. [d] Target year: 2000. [e] Source [45]

Table 3.16 Normalized results (milliperson equivalents)

Impact category	Plastic bag (HDPE)	Paper bag (kraft paper)	PP fibre nonwoven bag	Woven cotton bag	Boutique plastic (LDPE)
GWP 100 Years-kg CO_2 eq/person/year	1.37E−05	4.64E−05	4.02E−04	5.79E−04	5.13E−05
Ozone depletion potential—kg CFC11 eq/person/year	2.39E−05	8.47E−05	6.74E−04	1.06E−03	8.91E−05
Acidification—kg SO_2-eq/person/year	3.56E−05	2.02E−04	1.01E−03	1.59E−03	1.33E−04
Photo-oxidant chemical formation—kg C_2H_4-eq/person/year	9.91E−05	3.51E−04	2.79E−03	4.40E−03	3.69E−04
Radioactive solid waste—kg hazard. waste/person/year	1.64E−12	5.83E−12	4.62E−11	7.28E−11	6.11E−12
Slag and ash—kg slag and ashes/person/year	5.19E−06	1.84E−05	1.46E−04	2.30E−04	1.93E−05
ADP Chinese factors	4.89E−03	1.73E−02	1.38E−01	2.17E−01	1.82E−02
ADP-Chinese factors for norm—alisation figure of 4.67×10^{-11}	2.51E+03	8.91E+03	6.94E+04	1.12E+05	9.42E+03

Table 3.17 Unnormalised results

Impact category	Plastic bag (HDPE)	Paper bag (kraft paper)	PP fibre nonwoven bag	Woven cotton bag	Boutique plastic (LDPE)
Human toxicity	6.86E−02	2.45E−01	1.96E+00	3.08E+00	2.59E−01
Eutrophication	7.49E−05	2.66E−04	2.11E−03	3.32E−03	2.79E−04
Ionising radiation	6.55E−11	2.32E−10	1.84E−09	2.90E−09	2.44E−10
Depletion of abiotic resources	7.10E−04	2.52E−03	2.00E−02	3.15E−02	3.19E−03
Environmental burden-ELU-emissions	7.69E−03	2.61E−02	2.26E−01	3.26E−01	2.88E−02
Environmental burden-ELU-resources	2.33E−02	8.26E−02	6.57E−01	1.03E+00	8.69E−02
Emissions of waste water	1.44E−01	5.11E−01	4.06E+00	6.39E+00	5.37E−01

3.5.3 Verification of Results with SIMAPRO 7.2

The results from the manual calculation were verified by the results of SIMAPRO 7.2. Results of characterization were compared, since the equation and the values for characterization are common. The results are shown in Figs. 3.9, 3.10, 3.11, 3.12, 3.13, 3.14, 3.15, which show the correlation of manually calculated and software generated results.

Table 3.18 Weighed results (milli person equivalents)

Impact category	Plastic bag (HDPE)	Paper bag (kraft paper)	PP fibre nonwoven bag	Woven cotton bag	Boutique plastic (LDPE)
GWP 100 years	1.14E−05	3.85E−05	3.34E−04	4.81E−04	4.26E−05
Ozone depletion potential	6.45E−05	2.29E−04	1.82E−03	2.86E−03	2.00E−04
Acidification	2.60E−05	1.47E−04	7.35E−04	1.16E−03	9.73E−05
Photo-oxidant chemical formation	3.96E−05	1.41E−04	1.12E−03	1.76E−03	1.48E−04
Radioactive solid waste	7.38E−13	2.63E−12	2.08E−11	3.28E−11	2.75E−12
Slag and ash	1.27E−06	4.49E−06	3.57E−05	5.62E−05	4.72E−06

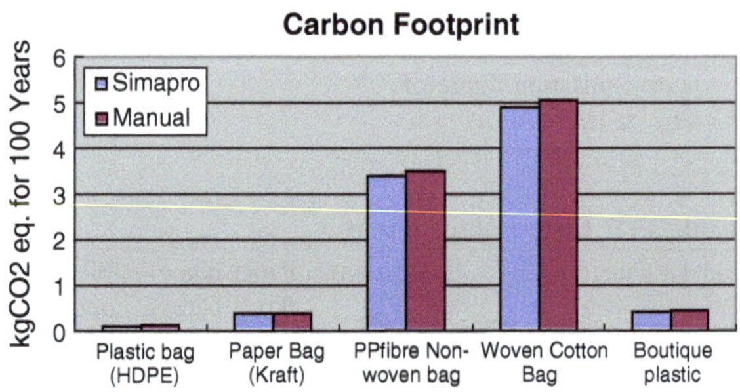

Fig. 3.9 Carbon footprint (results rounded-off) [24]

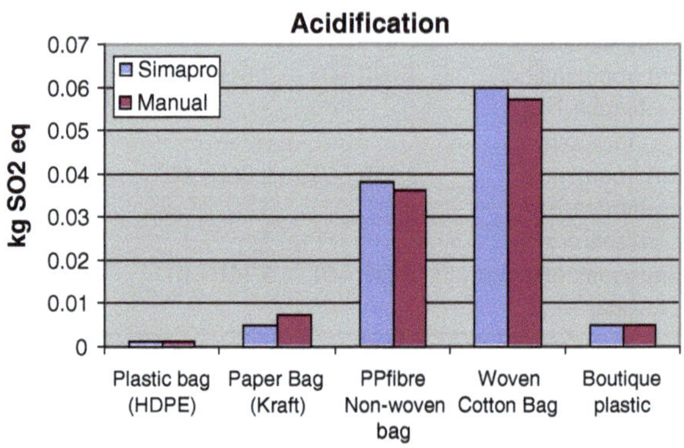

Fig. 3.10 Acidification (results rounded-off) [24]

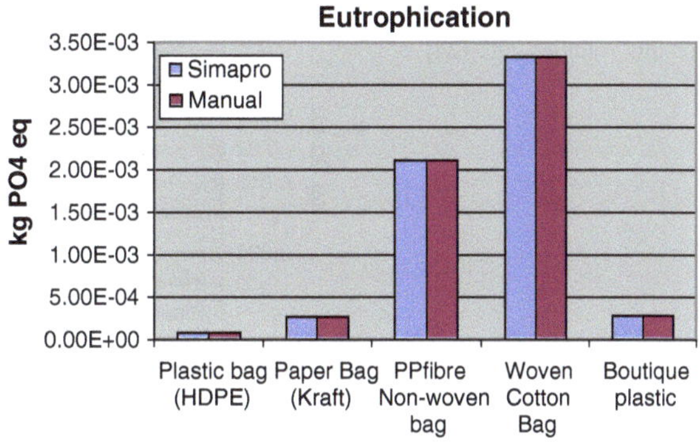

Fig. 3.11 Eutrophication (results rounded-off) [24]

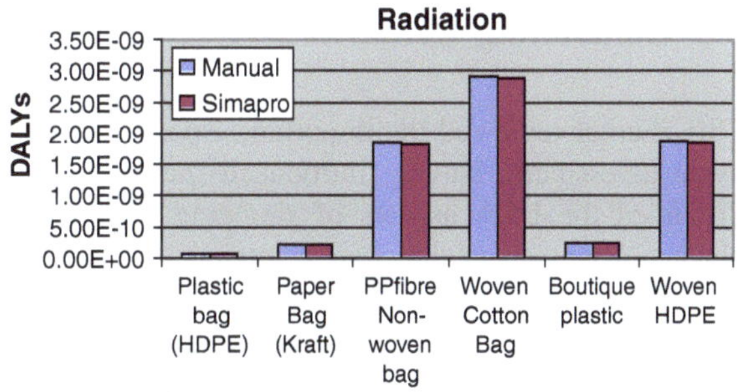

Fig. 3.12 Radiation (results rounded-off) [24]

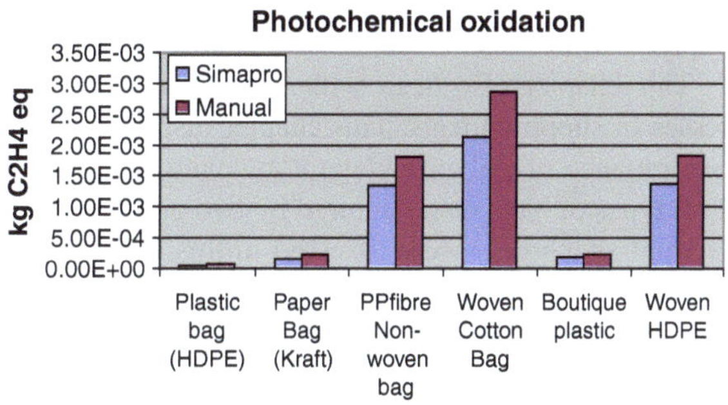

Fig. 3.13 Photo-oxidant chemical formation (results rounded-off) [24]

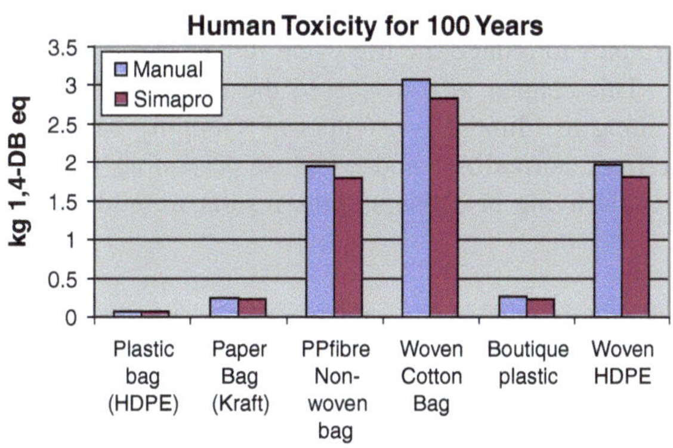

Fig. 3.14 Human toxicity (results rounded-off) [24]

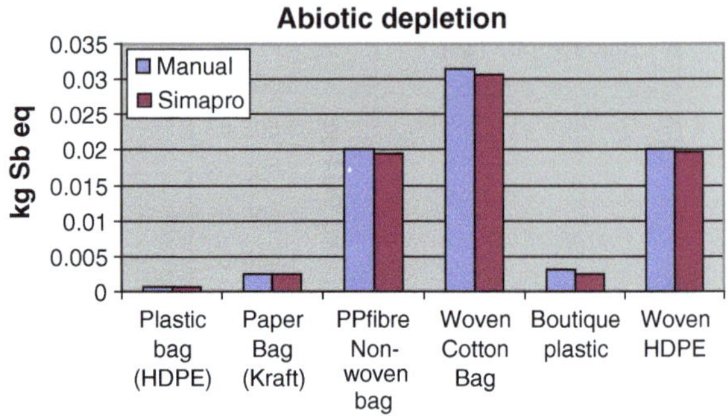

Fig. 3.15 Abiotic depletion (results rounded-off) [24]

3.6 Concluding Remarks

This chapter reviewed the important aspect in eco-fucntional assessment, i.e. life cycle assessment. Having introduced the concept of LCA briefly, this chapter discussed the LCA aspects of shopping bags used for grocery purposes. This chapter discussed an innovative attempt towards quantifying the environmental impact and ecological sustainability of different raw materials being used for making grocery shopping bags with the aid of a scientific model developed. According to this model, organic cotton followed by paper raw material top the entire list of raw materials used for shopping bags in terms of last environmental impact and better ecological sustainability.

This chapter also dealt with the life cycle assessment of manufacturing processes of shopping bags. This chapter discussed in detail about the environmental performance of polypropylene nonwoven bags in terms of life cycle assessment. Two types of bags manufactured by two production techniques namely sewing and thermal attachment were assessed in this study by the primary data collected on-site in a nonwoven manufacturing factory in southern china. Eco-damage by Eco-indicator'99 method and global arming potential by IPCC 2007 methods were used to quantify the environmental impacts made by these products. From the analysis it was understood that sewing technology seems to be better in terms of better environmental performance than its counterpart. Recommendations on feasible grounds to reduce the life cycle impacts were also discussed.

This chapter also discussed the manual calculation of life cycle impacts pertaining to Chinese consumers of shopping bags. Different impacts were characterized, normalized and weighed pertaining to the impact values applicable to people living in China and the results were also compared with software results.

References

1. Society of Environmental Toxicology and Chemistry (SETAC) (1993) Guidelines for life cycle assessment: a code of practice, First edn. From the SETAC workshop held at Sesimbra, Portugal, 31 March–3 April 1993. Brussels, Belgium, and Pensacola, Florida, USA
2. Fava JA, Denison R, Jones B, Curran MA, Vigon B, Selke S, Barnum J (eds) (1991) A technical framework for life cycle assessment. August 18–23, Smugglers Notch, Vermont, Society of environmental toxicology and chemistry (SETAC). SETAC, Washington, DC
3. ISO 14040 (2006) Environmental management-life cycle assessment-principles and frame work. International organisation for standardisation, Geneva
4. ISO 14044 (2006) Environmental management -life cycle assessment- international organisation for standardisation, Geneva
5. Muthu SS, Li Y, Hu JY, Mok PY (2012) Quantification of environmental impact and ecological sustainability for textile fibres. Ecol Ind 13(1):66–74
6. Photosynthesis, Available at: http//en.wikipedia.org/wiki/Photosynthesis
7. Jordan AG Facts about cotton and global warming, www.cottoninc.com/Air...Quality/Cotton-and-Global-Warming-Facts/
8. Mankowski J, Kolodziej J (2008) Increasing heat of combustion of briquettes made of hemp shives. In: International conference on flax and other bast plants. pp 344–352
9. Benefits of Trees in Urban Areas. http//www.coloradotrees.org/benefits.htm
10. Flatulent sheep cause global warming. http//www.ananova.com/news/story/sm_101581.html?nav_src=newsIndexHeadline
11. Renewable Resource. http://en.wikipedia.org/wiki/Renewable_resource
12. Chen H-L, Burns LD (2006) Environmental analysis of textile products. Clothing Text Res J 24(3):248–261
13. Richardson S, Gorton L (2003) Characterisation of the substituent distribution in starch and cellulose derivatives. Anal Chim Acta 497:27
14. Horrocks AR, Hall ME, Roberts D (1997) Environmental consequences of using flame-retardant textiles—a simple life cycle analytical model. Fire Mater 21(5):229–234
15. Chung HP, Yun KK, Seung SI (2004) Biodegradability of cellulose fabrics. J Appl Polymer Sci 94(1):248–253
16. Kaillala ME, Nousiainen P (1999) Environmental profile of cotton and polyester-cotton fabrics. AUTEX Res J 1(1):8–20
17. What is the energy profile of the textile industry? Available at: http//oecotextiles.wordpress.com/2009/06/16/what-is-the-energy-profile-of-the-textile-industry
18. Barber A, Pellow G (2006) Life cycle assessment New Zealand merino industry merino wool total energy use and carbon dioxide emissions. The Agri Business Group
19. Boustead I (2005) Eco-profiles of the European plastics industry. Plastics Europe, March 2005
20. Laursen SE, Hansen J, Bagh J, Jensen OK, Werther I (1997) Environmental assessment of textiles. Life cycle screening of the production of textiles containing cotton, wool, viscose, polyester or acrylic fibres. Environmental project no. 369. Ministry of the Environment and Energy, Danish Environmental Protection Agency
21. Morris D The Fibres, Textile and Textile Manufacturing Industries in China P.R Forecasts and Environmental Considerations, 77th International Wool Textile Organisation congress, Beijing, China, Obtained from: www.cirfs.org
22. Cherett N et al (2005) Ecological footprint and water analysis of cotton, hemp and polyester, Stockholm environment institute
23. Going Carbon Neutral: A Guide for Publishers, Accessed from: http://www.newsociety.com/Publishers%20CO2%20Template%20Metric.pdf. Accessed 15 Nov 2010
24. Muthu SS (2011) Eco-functional assessment of grocery shopping bags, Ph.D. thesis, The Hong Kong Polytechnic University, Hong Kong

25. Chaffee C, Yaros BR (2007) Life cycle assessment for three types of grocery bags—recyclable plastic; compostable, biodegradable plastic; and recycled, recyclable paper. Final report, Boustead Consulting and Associates Ltd
26. Ecobilan; Evaluation study of environmental impacts of paper and plastic carrier bags, 2008. Available at: http://www.eurosac.org/eurosac/pdf/3808_LCA-Carrier-Bags-Methodology-(english).pdf
27. Nolan ITU (2002) Centre for Design (RMIT), and Eunomia Research and Consulting Ltd; plastic shopping bags—analysis of levies and environmental impacts. Environmental Australia, Department of the Environment and Heritage, Commonwealth Government, Canberra
28. ExcelPlas Australia (2004) Centre for Design (RMIT), and Nolan ITU; the impacts of degradable plastic bags in Australia, final report to Department of the Environment and Heritage, Commonwealth Government of Australia, Canberra
29. James K, Grant T (2005) LCA of degradable plastic bags. In: 4th Australian LCA conference, Sydney, February 2005
30. FRIDGE. Socio-economic Impact Assessment of the Proposed Plastic Bag Regulations Report for the National Economic Development and Labour Council. Bentley West Management Consultants, Johannesburg, South Africa. Available at: sfenvironment.org/downloads/library/asticlifecycleanalysis.pdf
31. GUA—Gesellschaft für umfassende Analysen (2005) The contribution of plastic products to resource efficiency, Vienna. http://www.plasticseurope.org/Content/Default.asp?PageID=517
32. Los Angeles County Department of Public Works; an overview of carryout bags in Los Angeles County, August 2007
33. Paper vs. plastic bags. Downloaded at: http://iere.org/ILEA/lcas/franklin1990.html, 2004
34. The ULS Report (2008) Review of life cycle data relating to disposable, compostable, biodegradable and reusable grocery bags, 28 March 2008
35. Ellis S, Kantner S, Saab A, Watson M (2005) Plastic grocery bags: the ecological footprint, Student publications, VIPIRG publications, University of Victoria, PO Box 3050 STN CSC, Victoria
36. Dilli R (2007) Comparison of existing life cycle analysis of shopping bag alternatives, final report, Hyder Consulting Pty Ltd, Australia, April 2007
37. Muthu SS, Li Y, Hu JY, Mok PY (2010) An exploratory comparative study on eco-impact of paper and plastic bags. J Fibre Bioeng Informatics 4:307–320
38. Muthu SS, Li Y, Hu JY, Mok PY, Ding X (2012) Influence of consumer behaviour and governmental policies in China, Hong Kong and India: an eco-impact assessment study of reusable shopping bags EEST Part A. Energy Sci Res 28:1131–1144
39. Muthu SS, Li Y, Hu J, Mok PY, Liao X (2012) Carbon footprint of production processes of polypropylene nonwoven shopping bags. Fibres Text Eastern Europe 20, 3(92):12–15
40. Di XH, Nie ZR, Yuan BR, Zuo TY (2007) Life cycle inventory for electricity generation in China. Int J Life Cycle Assess 12(4):217–224
41. Guinée JB, Gorrée M, Heijungs R, Huppes G, Kleijn R, de Koning A; van Oers L, Wegener Sleeswijk A, Suh S, Udo de Haes HA, de Bruijn H, van Duin R, Huijbregts MAJ (2002) Handbook on life cycle assessment. Operational guide to the ISO standards. I: LCA in perspective. IIa: Guide. IIb: operational annex. III: Scientific background. Kluwer Academic Publishers, ISBN 1-4020-0228-9, Dordrecht, pp 692
42. Swerea IVF (2009) Research institute. ELU values for air pollutants. 10 March 2009. Accessed from: http://lotsen.ivf.se/KonsLotsen/Bok/Kap3/Miljopaverkan.html
43. Feng G, ZuoRen N, ZhiHong W, XianZheng G, TieYong Z (2009) Characterization and normalization factors of abiotic resource depletion for life cycle impact assessment in China. Sci China Ser E-Tech Sci 52(1):215–222
44. Yang J, Nielsen PH (2001) Chinese life cycle impact assessment factors. J Environ Sci 13(2):205–209
45. Wenzel H, Hauschild MZ, Alting L (1997) Environmental assessment of products. Methodology, tools, techniques and case studies in product development, vol 1. Chapman & Hall, United Kingdom, ISBN 0-412-80800-5

Chapter 4
Assessment of Functional Aspects of Shopping Bags

4.1 Introduction

Functionality of a product is the key factor in deciding its performance for the application/s for which it is being intended. Functionality plays a vital role in life cycle assessment or environmental impact considerations, since functionality is the base on which the useful life of a product is decided. The functional properties/aspects of a product can be defined as the properties related to the function of that particular product for which it is intended.

The proposed concept of Eco-functional Assessment is coined from the combination of Ecological and Functional aspects of any product, where the functionality assumes equal significance as ecological properties. Functional aspects differ from product to product. Every product has its unique considerations in terms of Functional aspects and Functional limits. Functionality of a product covers various aspects associated with that product such as physical, chemical, mechanical, aesthetic and surface.

This chapter deals with the various Functional aspects applicable to grocery shopping bags, test methods and standards as well as assessment of functional aspects of shopping bags and results. This chapter also deals with the reusability assessment of shopping bags with the aid of eco-functional tester developed by the authors.

4.2 Different Functional Aspects of Shopping Bags

There are many properties which decide the functionality of shopping bags. A functionality assessment must necessarily test all of them. Major functional properties which need to be considered in evaluating the functionality of grocery shopping bags are described below.

S. S. Muthu and Y. Li, *Assessment of Environmental Impact by Grocery Shopping Bags*, EcoProduction, DOI: 10.1007/978-981-4560-20-7_4,

4.2.1 Material Composition

Firstly, material composition, which deals with the assessment of type of fibre or raw material used in manufacturing a product under question. Composition may consist of a single entity or a blended one. A wide range of fibres or raw materials and combination of fibres or raw materials are generally employed to manufacture shopping bags and it is very much essential to identify the composition of a particular shopping bag product. It is also a legal requirement for the manufacturer to identify and label a material's composition. This is also termed "Fibre identification"/"Fibre content or composition determination"/"Fabric composition determination". There are many existing methods to identify fibre/material composition and all help us to identify the type of fibre in the material being tested [1, 2]:

1. Microscopical examination of the longitudinal and cross-sectional views of the fibre (Optical Test)
2. Burning test
3. The use of solvents and other chemical tests
4. Staining test
5. Fibre density
6. Miscellaneous methods such as melting point determination

These methods, listed above can determine the composition of single layer fabrics/materials. Traditional methods listed above will aid in finding out the fibre content alone, which alone will not be sufficient to cope with the pace of green consumerism. Test methods for fibre composition must also be helpful to analyse the products from the perspective of environmental impact assessment as well as by incorporating eco-testing features to analyse various elements such as banned azo colourants, formaldehyde content, heavy metal residues, ozone-depleting chemicals, pesticide residues [2]. These requirements stipulate new methods for material composition tests which are discussed below. Based on these requirements, many new methods have been developed for testing fibre composition, including, but not limited to:

1. Environmental scanning electron microscope (ESEM) technology
2. Near infrared spectral image measurement system
3. Capillary electrophoresis/mass spectrometry (CE/MS) technique
4. Thermogravimetry (TG) analysis
5. Computer image processing technology [2].

4.2.1.1 Standards for Material Composition Test

Many standards have also been developed for the material composition test. Some of the familiar and widely adopted ones include:

1. AATCC Test Method 20A-2007: Fibre Analysis: Quantitative
2. AATCC Test Method 20-2007: Fibre Analysis: Qualitative
3. ASTM D276: Standard Test Methods for Identification of Fibers in Textiles
4. ISO 1833: Textiles—Quantitative chemical analysis

   (a) ISO 1833-1:2006—Part 1: General principles of testing
   (b) ISO 1833-3:2006—Part 2: Ternary fibre mixtures
   (c) ISO 1833-3:2006—Part 3: Mixtures of acetate and certain other fibres (method using acetone
   (d) ISO 1833-4:2006—Part 4: Mixtures of certain protein and certain other fibres (method using hypochlorite)
   (e) ISO 1833-5:2006—Part 5: Mixtures of viscose, cupro or modal and cotton fibres (method using sodium zincate)
   (f) ISO 1833-6:2007—Part 6: Mixtures of viscose or certain types of cupro or modal or lyocell and cotton fibres (method using formic acid and zinc chloride)
   (g) ISO 1833-7:2006—Part 7: Mixtures of polyamide and certain other fibres (method using formic acid)
   (h) ISO 1833-8:2006—Part 8: Mixtures of acetate and triacetate fibres (method using acetone)
   (i) ISO 1833-9:2006—Part 9: Mixtures of acetate and triacetate fibres (method using benzyl alcohol)
   (j) ISO 1833-10:2006—Part 10: Mixtures of triacetate or polylactide and certain other fibres (method using dichloromethane)
   (k) ISO 1833-11:2006—Part 11: Mixtures of cellulose and polyester fibres (method using sulfuric acid)
   (l) ISO 1833-12:2006—Part 12: Mixtures of acrylic, certain modacrylics, certain chlorofibres, certain elastanes and certain other fibres (method using dimethylformamide)
   (m) ISO 1833-13:2006—Part 13: Mixtures of certain chlorofibres and certain other fibres (method using carbon disulfide/acetone)
   (n) ISO 1833-14:2006—Part 14: Mixtures of acetate and certain chlorofibres (method using acetic acid)
   (o) ISO 1833-15:2006—Part 15: Mixtures of jute and certain animal fibres (method by determining nitrogen content)
   (p) ISO 1833-16:2006—Part 16: Mixtures of polypropylene fibres and certain other fibres (method using xylene)
   (q) ISO 1833-17:2006—Part 17: Mixtures of chlorofibres (homopolymers of vinyl chloride) and certain other fibres (method using sulfuric acid)
   (r) ISO 1833-18:2006—Part 18: Mixtures of silk and wool or hair (method using sulfuric acid)
   (s) ISO 1833-19:2006—Part 19: Mixtures of cellulose fibres and asbestos (method by heating)
   (t) ISO 1833-20:2009—Part 20: Mixtures of elastane and certain other fibres (method using dimethyl acetamide)

(u) ISO 1833-21:2006—Part 21: Mixtures of chlorofibres, certain modacrylics, certain elastanes, acetates, triacetates and certain other fibres (method using cyclohexanone).

4.2.2 Physical, Mechanical and Dimensional Properties

Followed by the material composition, the next important category of functional aspects is the functional properties of shopping bags. Functional properties include a long list of properties and test methods related to physical, dimensional and mechanical aspects. Many kinds of tests are available for testing functional properties which are given below, and not limited to:

1. Areal density;
2. Thickness;
3. Tensile strength;
4. Tear strength;
5. Bursting strength;
6. Permeability tests;
7. Colour fastness tests;
8. Water and oil proof tests.

The different testing standards, testing equipment and methods used for the above functional properties are described in Table 4.1.

4.2.3 Safety Properties

Safety properties stand next to physical and mechanical properties in the functional properties list. There are many tests, that can be conducted to assess the human safety properties in different products. Some of the safety tests can be conducted in grocery shopping bags are given below, and not limited to:

1. pH;
2. Formaldehyde;
3. Forbidden Azo-benzene colouring matter;
4. Flammability;
5. Non-toxicity;
6. Anti-static;
7. Heavy metals.

The different testing standards, testing equipment and methods used for measuring human safety are listed in Table 4.2.

Table 4.1 Functional properties test methods & standards for shopping bags

Name of the test parameter	Testing equipment/s used to test	Standard/s applicable to test
Areal density	Weighing balance	• ASTM D3776/D3776 M-09a Standard test methods for mass per unit area (weight) of fabric • ISO 3801-1977—Textiles—Woven fabrics—Determination of mass per unit length and mass per unit area • ISO 9073-1:1989: Textiles—Test methods for nonwovens—Part 1: Determination of mass per unit area
Thickness	Thickness tester	• ISO 9073-2—Textiles. Test methods for nonwovens. Part 2: determination of thickness • ASTM D 1777-96—Standard Test Method for Thickness of Textile Materials • ASTM D 5729-97—Standard test method for thickness of Nonwoven fabrics • ISO 5084-1996—Textiles—Determination of thickness of textiles and textile products
Tensile strength and elongation	Tensile testing machines of CRE/CRL/CRT principles	• ISO 13934-1:1999 Textiles: Tensile properties of fabrics- Part 1: Determination of maximum force and elongation at maximum force using the strip method • ISO 13934-2:1999 Textiles: Tensile properties of fabrics—Part 2: Determination of maximum force using the grab method • ASTM D5034-09 Standard test method for breaking strength and elongation of textile fabrics (grab test) • ASTM D5035-06(2008)e 1 standard test method for breaking force and elongation of textile fabrics (strip method) • ISO 9073-3:1989: Textiles—Test methods for nonwovens—Part 3: Determination of tensile strength and elongation

(continued)

Table 4.1 (continued)

Name of the test parameter	Testing equipment/s used to test	Standard/s applicable to test
Tear strength	Elmendorf tester/Tensile tester (CRE) depending upon the testing standard	• ISO 4674-1998, part 1: Determination of tear resistance • ISO 13937-3-2000 Textiles—Tear properties of fabrics: Part 3: Determination of tear force of wing-shaped test specimens • ISO 13937-1-2000 Textiles—Tear properties of fabrics : Part 1: Determination of tear force using the ballistic pendulum method (Elmendorf) • BS 3424 Method 7C, Single tear, 1973 • EN 1875-3 Determination of tear resistance: Part 3: Trapezoid tear, 1997 • ASTM D1423-83 Tear resistance of woven fabrics by falling pendulum (Elmendorf) • ASTM D751 Tack tear, 1995 • ASTM D751 Puncture resistance, 1995 • ISO 5473 Determination of crush resistance, 1997 • ASTM D 5734 standard test method for tearing strength of Nonwoven fabrics by falling-pendulum (Elmendorf) apparatus • ASTM D 5735 standard test method for tearing strength on nonwoven fabrics by the tongue (single rip) procedure (constant-rate-of-extension tensile testing machine) • ISO 9073-4:1997—Textiles—Test methods for nonwovens—Part 4: determination of tear resistance
Bursting strength	Ball burst tester/Diaphragm bursting strength tester	• ISO 3303-1995 Determination of bursting strength • ISO 2960 Textiles—Determination of bursting strength and bursting distension—Diaphragm method • BS 4768 Method for determination of the bursting strength and bursting distension of fabrics • BS 3424 Methods of test for coated fabrics—Wounded burst test • ASTM D3787 Standard test method for bursting strength of knitted goods—constant-rate-of-traverse (CRT) ball burst test • ASTM D3786/D3786 M-09 Standard Test Method for Bursting Strength of Textile Fabrics—Diaphragm Bursting Strength Tester Method • ISO 13938-2:1999—Textiles—Bursting properties of fabrics—Part 2: Pneumatic method for determination of bursting strength and bursting distension

(continued)

Table 4.1 (continued)

Name of the test parameter	Testing equipment/s used to test	Standard/s applicable to test
Seam strength and slippage	Tensile tester of CRE type	• ASTM D1683 Standard test method for failure in sewn seams of woven fabrics, 1990 • ASTM D751 Seam strength, 1995 • BS 3320:1988 Method for determination of slippage resistance of yarns in woven fabrics: Seam method
Permeability tests	1. Air permeability tester	• ASTM D737-04 (2008) Standard test method for air permeability of textile fabrics • ISO 9073-15:2007: Textiles—Test methods for nonwovens—Part 15: Determination of air permeability • ISO 9237:1995—Textiles—Determination of the permeability of fabrics to air
	2. Water vapour permeability tester	• ASTM E96-00 • ASTM D6701-01 Standard test method for determining water vapour transmission rates through nonwoven and plastic barriers (withdrawn)
	3. Water repellency and water resistance testers (Impact penetration tester/Spray Tester/ Bundesmann rain tester/ Hydrostatic Pressure Tester)	• AATCC test method 70-2000 Water repellency: Tumble jar dynamic absorption test • AATCC Test method 22, Water repellency: spray test • AATCC Method 21, Water repellency: Static absorption test • ISO 18695:2007: Textiles—Determination of resistance to water penetration—Impact penetration test • ISO 18696:2006: Textiles—Determination of resistance to water absorption—Tumble-jar absorption test • ISO 9865:1991: Textiles—Determination of water repellency of fabrics by the Bundesmann rain-shower test • ISO 9073-17:2008: Textiles—Test methods for nonwovens—Part 17: Determination of water penetration (spray impact) • ISO 4920:1981: Textiles—Determination of resistance to surface wetting (spray test) of fabrics • AATCC Test method 127-2008: Water resistance: hydrostatic pressure test • ISO 811:1981: Textile fabrics—Determination of resistance to water penetration—Hydrostatic pressure test • ISO 22958:2005: Textiles—Water resistance—Rain tests: exposure to a horizontal water spray. • AATCC Test method 35, Water resistance: rain test • AATCC Method 42, Water resistance: Impact penetration test

(continued)

Table 4.1 (continued)

Name of the test parameter	Testing equipment/s used to test	Standard/s applicable to test
	4. MMT	• AATCC Test Method 195-2009—Liquid moisture management properties of textile fabrics
Colour fastness tests	1. Light fastness tester	• ISO 105-B01:1994—Textiles—Tests for colour fastness—Part B01: Colour fastness to light: Daylight • ISO 105-B02:1994—Textiles—Tests for colour fastness—Part B02: Colour fastness to artificial light: Xenon arc lamp fading test • ISO 105-B06:1998—Textiles—Tests for colour fastness—Part B06: Colour fastness and ageing to artificial light at high temperatures: Xenon arc lamp fading test • AATCC 16-2004
	2. Crockmeter (colour fastness to crocking)	• ISO 105-X12, • AATCC 8 & AATCC 165
	3. Laundr-o-meter (colour fastness to washing)	• ISO 105- C 06
	4. Glass plates, etc. (sea and chlorinated water fastness)	• AATCC 106:2007 & ISO 105-E02 :1996 (Colourfastness to Water: Sea) • AATCC 107:2007 & ISO 105-E01:1996 (Colourfastness to Water) • ISO 105 E03 & AATCC 162 (Chlorinated Water)
	5. Perspirometer (Perspiration fastness)	• AATCC 15-2007 • ISO 105-E04:1996
	6. Commercial Launderer (Domestic and commercial laundering fastness)	• ISO 105- C06/C08 • AATCC 61 No. 1A-5A
Oil Proof		AATCC 118

Table 4.2 Human safety—test methods & standards

Name of the test parameter	Standard/s applicable to test
pH	• ISO 3071:2005—Textiles—Determination of pH of aqueous extract • AATCC Test Method 81-2006—pH of the water-extract from wet processed textiles
Formaldehyde	• AATCC Test method 112-2008—Formaldehyde release from fabric, determination of: sealed Jar method • ISO 14184-1:1998—Textiles—Determination of formaldehyde—Part 1: Free and hydrolized formaldehyde (water extraction method) • ISO/DIS 14184-2—Textiles—Determination of formaldehyde—Part 2: Released formaldehyde (vapour absorption method)
Forbidden Azo-benzene colouring matter	• BS EN 14362-1:2003—Textiles. Methods for the determination of certain aromatic amines derived from azo colorants Detection of the use of certain azo colorants accessible without extraction • BS EN 14362-2:2003—Textiles. Methods for the determination of certain aromatic amines derived from azo colorants Detection of the use of certain azo colorants accessible by extracting the fibres
Flammability	• 16 C.F.R. Part 1610—Standard for the flammability of clothing textiles • ASTM D 6413—Standard test method for flame resistance of textiles (vertical test) • ASTM D1230—Standard test method for flammability of apparel textiles
Non toxicity	• OECD 201/202/203
Anti-static	• JIS L 1094
Heavy Metals	• EN 71 Part 3

4.3 Assessment of Functional Aspects of Shopping Bags

Eight types of shopping bags covering the majority of grocery shopping bag variety were chosen to study the Functional aspects (24). Plastic bags made out of Low and High Density Poly Ethylene (LDPE and HDPE), paper bags, nonwoven bags made out of polypropylene and polyester, manufactured from sewn and thermal technologies and cotton woven bags were selected for this study. Each type of bag was prepared specially for this study in three weight categories namely low, medium and heavy weight, based on the maximum possible weight that can be produced in each category. Bags made out of low weight are equivalent to 40 grams/square meter or whichever is possible in low weight category. Similarly medium and heavy weights are equivalent to 75 and 100 grams/square meter or whichever is possible in medium and heavy weight categories. In total 23 types of shopping bags were prepared under different weight categories as listed in Table 4.3. Table 4.3 lists the actual areal density of different samples and weight of 1 unit of bag.

Table 4.3 Details of weight ranges of samples selected for the study [3]

Sample number	Type of sample	Areal density (grams/sq.metre) (GSM)	Weight of one Unit of bag in grams.
1.	Paper 40 g	106.9	49.0
2.	Paper 75 g	132.4	58.3
3.	Paper 150 g	158.7	70.2
4.	Woven Cotton -1	188.1	118.5
5.	Woven Cotton -2	368.3	240.0
6.	HDPE -1	50.8	22.0
7.	HDPE -2	77.2	28.0
8.	HDPE -3	83.5	30.0
9.	LDPE -1	39.5	20.9
10.	LDPE -2	76.0	26.0
11.	LDPE -3	95.2	30.5
12.	PP 40 g Sewn	36.7	9.2
13.	PP 75 g Sewn	71.6	24.5
14.	PP 100 g Sewn	104.6	30.3
15.	PP 40 g Thermo	42.2	12.0
16.	PP 75 g Thermo	74.3	23.0
17.	PP 100 g Thermo	102.9	28.5
18.	PET 40 g Sewn	39.0	9.5
19.	PET 75 g Sewn	73.9	25.5
20.	PET 100 g Sewn	109.9	29.7
21.	PET 40 g Thermo	39.7	12.3
22.	PET 75 g Thermo	84.9	23.5
23.	PET 100 g Thermo	94.9	27.5

Different physical, mechanical, dimensional and other properties of shopping bags that decide the functionality of the bags were earmarked to assess the functionality, which are tabulated in Table 4.4.

The average result for each sample in the bunch of 23 samples selected for this study is shown in Figs. 4.1, 4.2, 4.3 and explained below category wise.

4.3.1 Strength Properties

As per Table 4.4, each sample was tested for tensile, tear and bursting strengths. Key parameters in each category such as maximum load in Newton for tensile strength, tear strength in Newton and bursting pressure sustained in pound per square inch (PSI) are expressed in Fig. 4.1.

From Fig. 4.1, it can be understood that cotton woven bags recorded maximum tensile strength amongst all samples selected for this study. Among LDPE and HDPE, HDPE surpassed LDPE in all weight categories in terms of tensile strength.

Table 4.4 Description of tested parameters and methods for functional properties [3]

Serial Number	Test parameter	Testing Standard employed	Testing machine employed	Testing parameters defined	No. of samples tested
1.	Tensile strength	ASTM D 5034(Grab Test) (ASTM D5034-09)	Instron (CRE)	300 mm/min Speed	5
2.	Tear strength	ASTM D 1424 & 1922 (ASTM D1922-09; ASTM D1424-09)	Elmendorf tearing tester	Measuring ranges: 200 gf, 400 gf, 800 gf, 1600 gf, 3200 gf, 6400 gf. (selected according to type of sample)	5
3.	Bursting strength	ISO 13938-2 (ISO 13938-2:1999)	Pneumatic tester, Truburst Bursting strength tester (James H Heal & Co. Ltd, England)	Area-7.3 cm^2; Dia-30.5 mm	5
6.	Thickness	ASTM D1777 (ASTM D1777-96(2007)	SDL fabric thickness tester	Pressure-5 gf/cm^2	5
7.	Areal density	ASTM D3776 (ASTM D3776/D3776 M-09)	Balance		5
8.	Air permeability	ISO 9237 (ISO 9237:1995)	Air permeability tester	20 cm^2 surface area and 100 Pa pressure drop	5
9.	Water vapour permeability	ASTM E 96 (ASTM E96/E96 M-10)	As per the standard	As per the standard	3
10.	pH test	ISO 3071 (ISO 3071:2005	Stoppered glass and mechanical shaker and others	As per the standard	3
11.	Formaldehyde content	ISO 14184-1 (ISO 14184-1:1998)			3
12.	Colour fastness to light	ISO 105 B02 (ISO 105-B02:1994)	Xenon arc lamp apparatus	BWS 4	Tested with BWS

(continued)

Table 4.4 (continued)

Serial Number	Test parameter	Testing Standard employed	Testing machine employed	Testing parameters defined	No. of samples tested
13.	Colour fastness to rubbing	ISO105 X12 (ISO 105-X12:2001)	Crockmeter	As per the standard	Dry and wet state
14.	Colour fastness to washing	ISO 105 C 06 -B2 (ISO 105-C06:2010)	Laundrometer	In laundro meter 30 min, Temp 40 °C, 25 steel balls, 4 gpl ECE phosphate +1 gpl Sodium perborate for 150 ml	Tested with multifibre
15.	Colour fastness to perspiration	ISO105 E 04 (ISO 105-E04:1994)	Test devices as advised by the Standard	4 h @ 37 °C, acid and alkaline conditions	Tested with multifibre
16.	Colour fastness to water	ISO 105 E 01 (ISO 105-E01:1994)			Tested with multifibre
17.	Fibre Content	ISO 1833-1 (ISO 1833-1:2006)	As per the standard	As per the standard	As per the standard

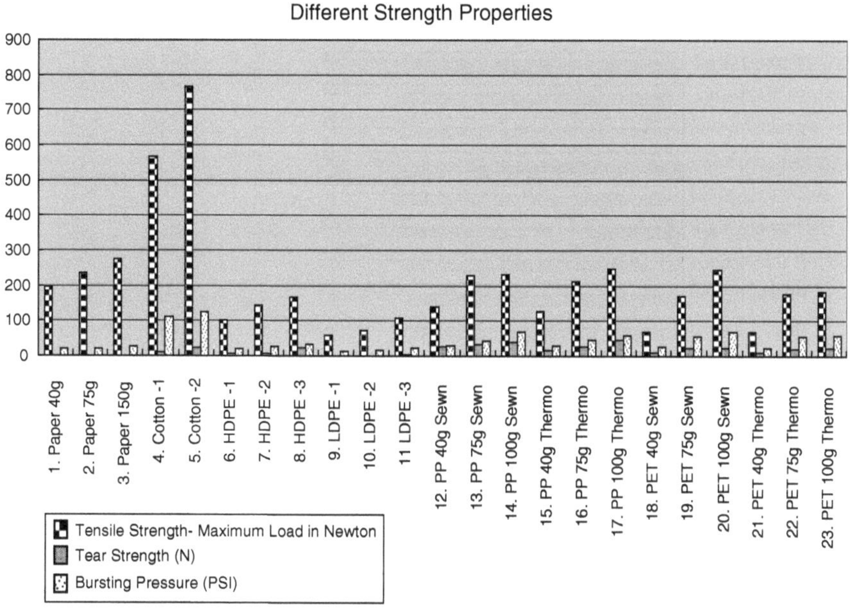

Fig. 4.1 Different strength properties of shopping bags [3]

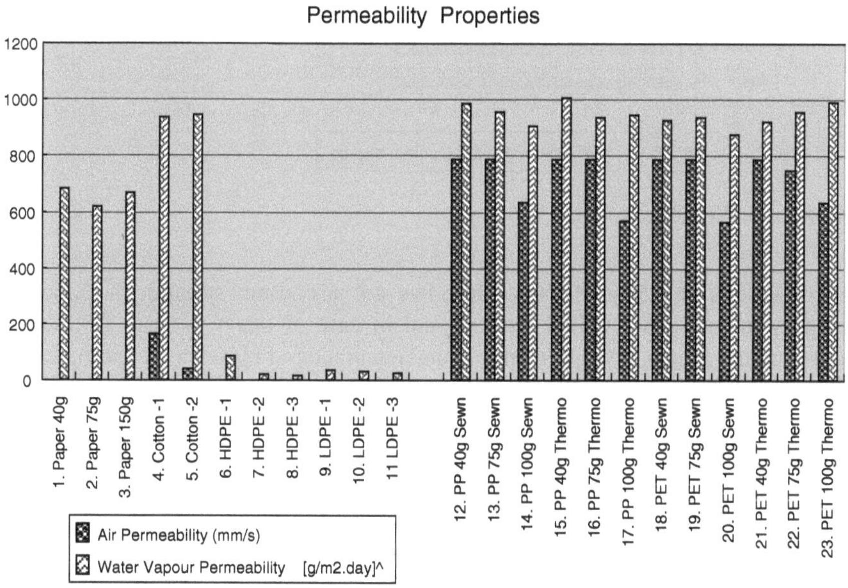

Fig. 4.2 Permeability test results of shopping bags [3]

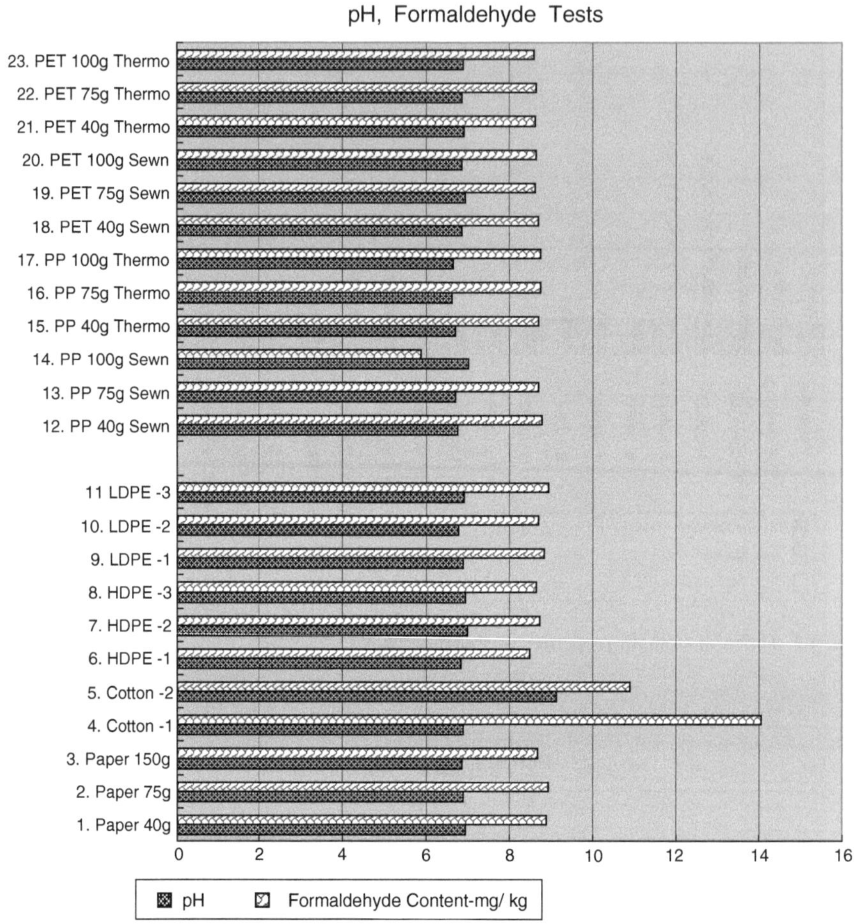

Fig. 4.3 pH, Formalehyde Tests [3]

In nonwovens category, PP sewn bag has got maximum strength than thermo category in low and medium weights and in case of heavy weight, PP thermo registered maximum strength than the one manufactured by sewn technology. The case is entirely reverse for PET bags, i.e. PET thermo has got maximum strength than sewn category in low and medium weights and in case of heavy weight, PET sewn registered maximum strength than the one manufactured by thermal technology.

As far as tear strength is concerned, PP bags of sewn technology have recorded maximum tear strength than its counterpart. Paper and LDPE categories of bags were found to have poor tear strength. In case of bursting strength, again, cotton bags have registered maximum bursting strength. Similar to tear strength, paper and LDPE possessed lower bursting strength.

Table 4.5 Thickness and Areal Density results [3]

Sample description	Thickness in mm.	Areal density in grams/sq. meter (GSM)	Fibre composition/content
1. Paper 40 g	0.08	106.9	100 % Paper
2. Paper 75 g	0.21	132.4	100 % Paper
3. Paper 150 g	0.24	158.7	100 % Paper
4. Cotton -1	0.54	188.1	100 % Cotton
5. Cotton -2	0.98	368.3	Cotton/Poly 52.5/47.5 (lining: 100 % cotton)
6. HDPE -1	0.10	50.8	100 % Polyethylene
7. HDPE -2	0.14	77.2	100 % Polyethylene
8. HDPE -3	0.18	83.5	100 % Polyethylene
9. LDPE -1	0.04	39.5	100 % Polyethylene
10. LDPE -2	0.09	76.0	100 % Polyethylene
11 LDPE -3	0.13	95.2	100 % Polyethylene
12. PP 40 g Sewn	0.35	36.7	100 % Polypropylene
13. PP 75 g Sewn	0.46	71.6	100 % Polypropylene
14. PP 100 g Sewn	0.61	104.6	100 % Polypropylene
15. PP 40 g Thermo	0.35	42.2	100 % Polypropylene
16. PP 75 g Thermo	0.55	74.3	100 % Polypropylene
17. PP 100 g Thermo	0.68	103.0	100 % Polypropylene
18. PET 40 g Sewn	0.30	39.0	100 % Polyester
19. PET 75 g Sewn	0.50	73.9	100 % Polyester
20. PET 100 g Sewn	0.55	109.9	100 % Polyester
21. PET 40 g Thermo	0.33	39.7	100 % Polyester
22. PET 75 g Thermo	0.52	84.8	100 % Polyester
23. PET 100 g Thermo	0.56	94.8	100 % Polyester

4.3.2 Dimensional Properties and Fibre Composition

Two type of dimensional attributes namely thickness and areal density and fibre composition were tested as per the standards and other specifications given in Table 4.4 and the tested results are tabulated in Table 4 5. As seen in Table 4.5, sample no. 5 was found to have cotton blended with polyester, in contrast to what was labeled. This is a critical issue affecting not only functional attributes, rather subjected to a lot of legal issues. This will be discussed in detail in Eco-functional assessment chapter.

4.3.3 Permeability Tests

Air and water vapor permeability characteristics of selected 23 samples were tested as per the details furnished in Table 4.4. Results of these parameters are shown in Fig. 4.2.

From Fig. 4.2, it can be understood that paper, LDPE and HDPE recorded lowest air permeability values and PP and PET nonwoven bags were found to have maximum air permeability. Again, PP and PET nonwoven bags registered maximum water vapour permeability also compared to all other categories of bags tested. LDPE and HDPE bags recorded lower amount of water vapour permeability.

4.3.4 Safety Tests

Two important parameters which portray safety aspect were tested, i.e. pH level and the presence/absence (if present, the amount) presence and amount of formaldehyde. Test results are displayed in Fig. 4.3.

All the samples tested above found to have very low levels of formaldehyde except cotton samples which have got slightly higher values compared to the other 21 samples. However, the result of formaldehyde test of all samples is in not detected range. For pH, except cotton-2 sample, all have got less than 7 pH (acidic), except one PP sample which recorded 7 value (neutral). Only one sample, i.e. cotton of heavy weight recorded alkaline pH.

4.3.5 Colour Fastness Tests

Ability to resist colour fastness against different agents namely light, water, acid and alkali perspiration, dry and wet rubbings were tested as per the specifications detailed in Table 4.4. Colour fastness results were up to the mark for all samples. All samples recorded fastness scale rating of 4 for light, washing (overall rating), acid and alkali perspiration (overall rating), water (overall rating). Cotton-1 and 2 were found to have 3/4 and 2/3 rating for body (dry and wet respectively) and 4/5 for lining material in it against dry and wet rubbing. All 6 PP samples recorded 4/5 grading for both dry and wet rubbing. For paper, HDPE, LDPE and PET samples rubbing fastness is not applicable [3].

4.4 Reusability Assessment: Development of Eco-Functional Tester

Along with the functional properties discussed above, there are another important set of properties, which we term here as eco-functional properties (since they decide Functional and ecological properties of shopping bags) which influence the functionality, consumption and disposal behaviors. A very important property, which lies at the interface of ecological and functional properties, is the reusability of shopping bags. Reusability of a shopping bag is a key factor which drives consumption behaviour primarily along with functionality and disposal aspects. Other relevant properties in the eco-functional properties list include impact strength and weight-holding capacity of grocery shopping bags. There is no equipment and methods developed so far to assess these pivotal characteristics of grocery shopping bags. Hence, an eco-functional tester was developed by authors and their team to assess these properties of grocery shopping bags [3, 4]. The developed tester can be utilized to assess these three properties (reusability, impact strength and weight-holding capacity) of any type of shopping bag. This section discusses the concept and development of an eco-functional tester for shopping bags. It also discusses the test results of the reusability, impact strength and weight-holding capacity of different types of shopping bags tested with the aid of developed tester. Authors and their team decided to express reusability and impact strength by two variants: absolute maximum capacity and comparative maximum capacity.

Grocery Shopping bags made out of any material, such as polyethylene, paper, cotton, polypropylene, jute, nylon, etc. and manufactured by any technique, nonwoven, woven, knitting technologies, plastic and paper bag processes, and so on, are primarily expected to be used many times for the benefit of economy and environment. The primary functions expected from a grocery shopping bag are given below, which decide the functionality of a bag and these functions decide the life span of a shopping bag:

1. How much time and how much weight, a shopping bag can sustain;
2. How many times can it be reused to carry a specified amount of weight, as per the claim of the manufacturer or in general according to the capacity of the bag;
3. How much impact can it withstand [3, 4].

3Rs has become a buzz word heard repeatedly worldwide in these days and it consists of three important terms: Reuse, Recycle and Reduce. Amongst these terms, the first and foremost one is, Reuse. This decides both the ecological and functional properties of a shopping bag. In simple terms, if a shopping bag is reused many times, due to its added functionality, its ecological benefits are included in terms of avoiding/postponing the depletion of resources and other pertinent environmental impacts to manufacture another bag. Again, preventing a shopping bag being recycled (though it is advantageous, it also demands resources, causes emissions and impacts) or sent to landfill earlier (which is a nightmare) also limits environmental damage and hence eco-damage could be prevented. Hence these important properties namely, reusability, weight-holding capacity and impact strength decide the eco-functional properties of shopping bags [3, 4].

It is very much significant to quantify these properties to assess the eco-functional characteristics of a shopping bag. As said earlier, there are no instruments available so far to assess eco-functional characteristics scientifically, which will aid both manufacturers and customers.

Results from this kind of instrument can be utilized for Life Cycle Assessment (LCA) studies, which deal with the quantification of the environmental impact made by any product/process in its useful time. The value of reusability obtained

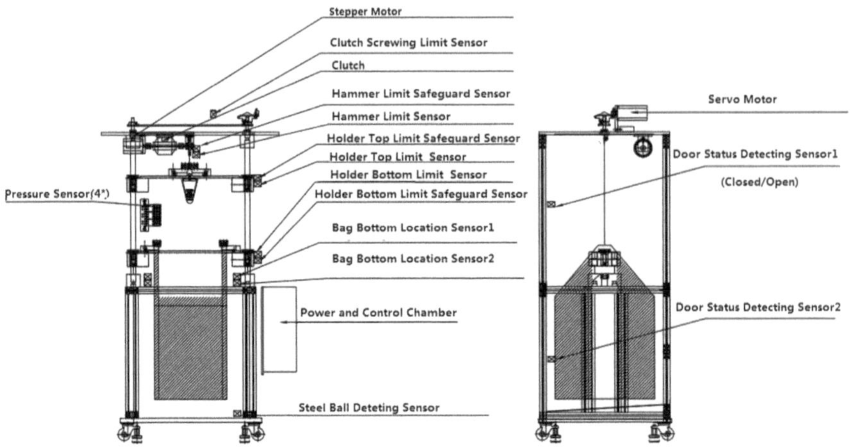

Fig. 4.4 Eco-functional Tester [3, 4]

from eco-functional tester can be directly utilized for LCA calculations. Other functions derived from this instrument are equally important, since they decide the useful life time of a shopping bag and they assist the LCA practitioner to decide the functional unit of the study. This unit is the base of any LCA study and upon which comparisons are made [3, 4].

As discussed earlier in previous chapter, a large number of studies have been conducted to investigate the LCA of various shopping bags and included an end-of-life assumption in the modelling. With the aid of the developed instrument under question, there is no need to assume the end-of-life values at least for the reusability function, which is one of the crucial considerations in end-of-life scenarios. Also with the aid of the developed instrument, it is possible to compare the actual reuse values of shopping bags derived from the developed instrument with the values derived from surveys of consumers (obtained from consumer behaviour studies, which will be discussed later in the next chapter). From the results of this equipment, it is possible to suggest consumers to modify their behaviour to prevent the environmental impact arising from the early disposal of shopping bags before their actual end-of-life [3, 4].

An eco-functional tester shown in Fig. 4.4 was developed to test the following properties of shopping bags, which will be assessed by this new instrument:

1. Reusability—Number of times a bag can be reused;
2. Holding Capacity of a bag-Amount of weight a bag can withstand for a designated time period;
3. Impact Strength—Assessment of strength of a bag when a sudden impact is applied [3, 4].

4.4.1 *Assessment of Eco-Functional Properties*

The working principle of the developed instrument is described below:

1. *Reusability*

Reusability will be tested by clamping the handle of the bag into the holding mechanism of the instrument and maintaining a designated load on to the bag under testing and subjecting it to a to and fro motion till the bag is broken. Specific number of times a bag withstands the designated load and able to sustain the motion will be recorded.

2. *Holding Capacity of a bag*

Holding capacity will be tested by clamping the handle of the bag into the holding mechanism and maintaining a particular load onto the bag while keeping it in a still state until it is broken. Alternatively, the bag is held for a certain period of time say 1 or 2 min or more, as required. Keeping time as constant, one can play with the maximum amount of load a bag bears for a fixed amount of time and that can be recorded.

3. *Impact Strength*

This will be tested by dropping designated weights, say 1, 2…5 kg, etc. into the bag from a certain distance (Since impact = mass*distance), thereby creating a progressive impact force that tears the shopping bag. Number of impact cycles a bag can withstand with a particular amount of load can be recorded.

All the required data can be acquired from the monitor installed in the instrument through PLC. This machine is equipped with 16 active sensors and two inactive ones. There are 6 load sensors, out of six, four sensors will sense the load carried by a bag. Information regarding the load carried by these four sensors will be transmitted to the PC and saved in Excel format and displayed as a graph. In case of a failure of the handles of a bag, one of the four load sensors will stop the machine due to the drop in the load. If there is a break in the body of the bag, the drop safety sensor shown in Fig. 4.4 will stop the machine.

4.4.2 *Experimentation of Eco-Functional Properties of Shopping Bags*

Selected 23 types of shopping bags were assessed for their eco-functional properties by the developed instrument. All the three functions were assessed for all 23 types of shopping bags.

4.4.2.1 Weight Holding Capacity

All the shopping bag samples selected for the study were tested for their maximum weight-holding capacity for 5 min. Within this selected 5-min period, samples were observed for any kind of failure such as hole formation, crack formation, propagation of tear/break in any part of the sample and failure of handles and body. For each sample, 3 specimens were tested. Three individual readings were taken and the averages of these three readings with error bars were reported. Each sample was tested from the lowest level of weight to the highest weight by gradually adding weights of 1 kg and the results at the maximum load for various shopping bags were reported [3, 4]. From the test results, it was understood that the cotton woven bags have the maximum weight-holding capacity in the list of chosen samples. They withstood up to a maximum of 35 kg and paper bags were found to have the lowest weight-holding capacity. It was found that even a heavy weight paper bag could only hold a maximum 8 kg for 5 min. LDPE and HDPE of various unit weights did record a maximum of 20 and 25 kg respectively. In case of thermo-bonded nonwoven bags made from PET of 40 GSM recorded a maximum of 12.3 kg and a sewn bag of same type withstood around 15 kg. But, PP thermo-bonded bags of the same weight withstood a maximum of 14 kg and survived at 20 kg if sewn. Nonwoven bags of 75 and 100 GSM, survived a maximum load of 25 kg without any failure, which is the maximum load they can hold for their sizes.

4.4.2.2 Reusability

Reusability was tested for two variants- absolute maximum capacity and comparative maximum capacity. To assess the reusability of shopping bags under absolute maximum capacity category, a fixed number of 100 cycles was chosen. Keeping the number of cycles constant, this test was conducted by varying the weights to determine the maximum reusability of a bag at different weight levels to establish the maximum weight carrying capacity of the bag. Paper bags, being the weakest of all samples selected here, recorded only 4 kg of weight even with their heaviest weight sample. Woven cotton bags can withstand 35 kg, which is the maximum load one call practically fill on to them. LDPE and HDPE bags can hold a maximum of 15 kg and their medium and heavy weight bags can withstand this load and complete 100 cycles.

Nonwoven PP bags in both the thermo and sewn categories withstood a maximum load of 15 kg for 100 cycles, while PET thermo bags fulfilled 0 and sewn bags fulfilled an average of 35 cycles at 15 kg of load. All nonwoven bags of 75 and 100 GSM withstood the maximum load of 20 kg for 100 cycles.

To quantify the comparative maximum capacity of selected shopping bags, samples were tested for 500 cycles at 10 kg. This load and number of cycles are referred from green seal standard-GS-16 [5]. Three samples were tested for each category and the average results were reported. From the results, it was noticed that except paper bags of all weight levels and PP 40 g sewn, PET 40 g thermo, LDPE-1, all other bags under discussion were capable of withstanding the prescribed load and able to complete the desired number of cycles. This is one of the very important parameters used directly for life cycle assessment simulation calculations [3, 4].

4.4.2.3 Impact Strength

a. Absolute Maximum Capacity

For this category of test, a fixed number of 5 cycles were chosen and the samples were tested at different loads between 2 and 5 kg. From the results, it was revealed that the paper, plastic and nonwoven bags of 40 GSM survived only 2 kg of load. The number of cycles they withstood with a load of 2 kg was recorded and reported. It was noticed that all paper, plastic and PET bags (sewn and thermo) of 40 g. survived just one cycle of impact load of 2 kg PP bags of 40 GSM (sewn and thermo) bore an average of 2.7 cycles of 2 kg impact load. Nonwoven bags of 75 and 100 GSM withstood only 3 kg of load and the number of cycles they withstood with the maximum load of 3 kg was recorded. Sewn PP bags of both 75 and 100 GSM, withstood a maximum of 5 cycles with 3 kg load and thermo PP bags of both 75 and 100 GSM endured only 3.7 cycles with 3 kg load. Woven bags survived the maximum load of 5 kg and the number of cycles they withstood with this load was recorded and reported [3, 4].

b. Comparative Maximum Capacity

To assess the comparative maximum capacity of shopping bags to determine their impact strength, samples were tested for the maximum number of cycles they could withstand with 5 kg. Three samples were tested for each category and the average results were reported. From the results, it was noticed that the nonwoven bags of 75 and 100 GSM and the woven bags are only able to fulfill the entire 5 cycles with the designated load of 3 kg [3, 4].

4.5 Concluding Remarks

This chapter dealt with an important aspect of eco-functional assessment, i.e. functionality aspect. Details of various functional aspects, test methods, standards applicable for grocery shopping bags were discussed in this chapter. 23 types of grocery shopping bags covering 8 major categories of nags were selected for this assessment. Different functional properties of shopping bags to assess the physical, mechanical and dimensional aspects were tested and the detailed test results were discussed. This chapter also dealt with another important aspect, i.e. reusability, impact strength and weight holding capacity of grocery shopping bags, which are pivotal to assess the ecological and functional properties of grocery bags (hence also termed as co-functional properties). An instrument named, co-functional tester was developed to assess these properties. The working principle, assessment of co-functional properties of selected grocery bags using this equipment as well as the results of this assessment were discussed in detail.

References

1. Taylor AM (1990) Technology of textile properties, Third edn. Forbes Publications, USA
2. Zhong Z, Xiao C (2008) Fabric composition testing in Fabric Testing. In: J Hu (ed) Woodhead textiles series no. 76. Woodhead publishing Limited, UK
3. Muthu SS (2011) Eco-Functional assessment of grocery shopping bags, PhD thesis, The Hong Kong Polytechnic University, Hong Kong
4. Muthu SS, Li Y, Hu JY, Mok PY, Mao YF, Wu XX, Li QH (2013) Assessment of eco-functional properties of shopping bags: development of a novel eco-functional tester. Int J Clothing Sci Tech 25(3):208–225
5. Green Seal (2008) GS-16 Reusable Bag Proposed Revised Standard. Green Seal Inc, Washington, D.C

Chapter 5
Consumption Behaviour of Shopping Bags and Eco-Impact

5.1 Introduction

Consumption behaviour plays a vital role in deciding the eco-impact made by consumer products. In the life cycle assessment, consumer behaviour plays a key role in deciding the potential impacts from gate to grave stages. Consumption behaviour lies purely in the hands of consumers and it is mainly driven by the functionality and ecological properties of a particular product. It is highly needed to study the consumer behaviour of any product and gather the real data from the actual consumers of products to be studied and incorporate the respective data in LCA calculations for use and disposal phases. This attempt will portray the real scenario, and it is highly expected than the one modelled by assumptions to fulfill the data for use and disposal phases. However, it is not an usual practice followed by LCA practioner's and it has not been reported in literature. The case is the same for grocery shopping bags and the authors have performed pioneering work in this area for grocery shopping bags and this chapter primarily deals with it.

Authors made an attempt to study the consumer behavior in terms of shopping bags used in Mainland China, Hong Kong and India among different user groups to deduce the end-of-life scenario values, such as the percentage of recycle, reuse and disposal to landfill of various shopping bags and these values were incorporated in life cycle assessment calculations to assess the eco-impact made by various grocery shopping bags. This issue is discussed in detail in this chapter.

5.2 Consumption Behaviour and its Influence on Environmental Impacts Made by Grocery Shopping Bags

A product's life cycle comprises of different phases namely raw material extraction, manufacturing phase, transportation, use and disposal phases. Out of which, use and disposal phases and their consequences on life cycle impacts are primarily decided by the consumers and this depends again on their behaviour or attitude

S. S. Muthu and Y. Li, *Assessment of Environmental Impact by Grocery Shopping Bags*, EcoProduction, DOI: 10.1007/978-981-4560-20-7_5,

towards a particular product. As said earlier, this attitude or behaviour is governed by the functionality and the ecological property of the discussed product. Similarly, the impacts of other phases namely raw material extraction, manufacturing, transportation are decided by the manufacturer of the particular raw material/ product or the person responsible for transportation.

The end-of-life or disposal phase is equally detrimental as the manufacturing phase for products like shopping bags which have limited shelf life (so to say, single use or quickly disposable products after very limited number of use). Products such as shopping bags in general do not have a sizable use phase which contributes significantly to life cycle impacts. After use phase, once the product is decided to be disposed of, it can be sent to recycling or incineration or sent to landfill or it can be reused. Human dimensions play a major role in consumer behaviour to decide upon a product's use and disposal phases and consequently its eco-impact or life cycle impacts. Apart from human dimensions, governmental policies also assume significant importance in the environmental impact made by a product, when it comes to the consideration of disposal options. Consumer behaviour and governmental policies go hand in hand and they jointly play a significant role in the disposal stage of shopping bags. Governmental policies influence the consumer behaviour also, apart from directly they rule upon the disposal options in terms of encouraging recycling policies, providing more recycling bins.

When it comes to shopping bags, use phase do not contribute much to life cycle impacts. At disposal stage, they can either be reused for primary or secondary purposes or sent to landfill or directed to recycling/incineration options. Incineration is not very common in Asia-pacific region. Out of all, reuse is the most environmentally beneficial option followed by recycling and the most detrimental one is landfilling. Consumer behaviour along with governmental policies govern the disposal option of shopping bags. Consumer behaviour influences the decision of use and disposal in the following manner: Despite the fact that certain types of bags are designed to possess sufficient functionality to be recycled and reused, it is left to the consumers to reuse a bag for the same purpose (primary reuse) until it can be discarded or recycled, i.e., to reuse the shopping bags many times until they can be disposed of. Upon deciding to dispose, they must be placed in recycling bins provided by the government or sent to recycling option, rather than disposed to landfill, which is a nightmare to the environment and has a corresponding eco-impact. Hence, consumer behaviour and governmental policies are the key factors in deciding the eco-impact of shopping bags.

5.3 Consumer Attitude/Behaviour in Various Countries towards the Use and Disposal of Grocery Shopping Bags

Consumer attitude or behaviour towards consumption and disposal of shopping bags varies between different individuals and it varies very significantly between various countries. Many factors influence the same such as the economical situation, laws prevailing in a particular country which influence consumption and disposal behaviour of shopping bags.

Many LCA studies conducted previously to compare the different grocery shopping bags in terms of their life cycle impacts have yielded important findings [1–13]. Some of the studies in this list included an end-of-life assumption to model the life cycle assessment of shopping bags; however this scenario is very far from reality. The reality of the end-of-life scenario lies primarily in the hands of consumers and the end-of-life scenario is consumer and country specific. It varies between different consumers in a same country and again, consumers of different countries also, hence generalization is not possible in this case. Under these circumstances, assumed end-of-life scenario would not work out and will not represent the reality.

None of the previous studies mentioned above, reported the use of real data arising from recycling/reuse/landfill options, obtained from the consumers of shopping bags, which prompted the researcher's interest. And hence the authors conducted a consumer survey in China, Hong Kong and India among different user groups of grocery shopping bags. Survey was administered through a questionnaire in these territories to understand the consumer's perception and behaviour towards different variety of grocery shopping bags [14]. This research work focused mainly on contemplating the consumer's perception and behavior on the usage and disposal of various shopping bags (frequency of reuse, recycle of different shopping bags and also their propensity to dispose these to landfills). This survey, focused on plastic, paper, nonwoven and woven bags, also investigated the existing policy dimensions of government on recycling phase and other associated factors related to it. Outcome of this survey enabled the authors to construct end-of-life scenarios in life cycle assessment by using real values from the actual users of shopping bags.

Having chosen with Convenience sampling method, this survey was answered by 100 respondents from China and 125 from India and Hong Kong, where the respondents were contacted by electronic means. The questionnaire used to administer this survey had 9 primary questions pertaining to the usage and disposal of shopping bags and 4 questions pertaining to the personal particulars of respondents. All the respondents fell in various age groups between 21–50 years, majority of them belong to 21–30 years in all the three territories and majority of the respondents are females in China and HK, whereas in case of India, majority of them are males. As far as the educational qualifications of respondents were concerned, majority were postgraduates in China and India, whereas majority were undergraduates in HK. This is a self acknowledged major limitation of this study, i.e. the demographic profile of the respondents was primarily biased towards the young and educated population. Nevertheless, conducting this study amongst the young and educated respondents could well indicate future trends in Asia without any doubt [14].

5.3.1 Reuse, Recycle, Disposal to Landfill Responses from Respondents

The primary aim of this survey was to collect the details of reuse, recycle and disposal options opted by various consumers in China, HK and India. Perceived reusability for primary or secondary uses were collected from various users from this study and from the results effective reuse percentage was calculated with the aid of the following two equations [14]:

$$\begin{aligned}\rho &= \text{Total Disposal }\% \\ &= P_0 * 100\,\% + P_1 * 100\,\% + P_2 * 50\,\% + P_3 * 25\,\% + P_4 * 10\,\%\end{aligned} \quad (5.1)$$

$$\text{Reuse }\% = 100 - \rho \quad (5.2)$$

where P_0 is the percentage of reuse for <1 time and so on. In the above Eq. 1, the corresponding percentage of <1 and 1 time usage is considered as 100 %, 2 times as 50 %, 3–5 times as 25 % and others (>5 times) is on an average considered as 10 %: Effective reuse % of different shopping bags calculated for the three territories are indicated in Table 5.1. Similarly from the perceived recycle options, an effective recycling amount in % was calculated by weighted average method for the three territories and are listed in Table 5.1.

Apart from this, customers were asked to fill in their opinion on how much percentage or how many times each bag can be reused/recycled/sent to landfill. This detail was filled by every user in all the territories and the average of the results are presented in Table 5.2.

Apart from this to understand the willingness of the respondents to place the used bags in recycling bins, questions related to the provision of recycling systems in their respective country were also asked and analysed [14]. From the analysis,

Table 5.1 Effective reuse and recycle percentages of grocery shopping bags

Type of bag	Reuse			Recycle		
	India	HK	China	India	HK	China
Plastic	28	38	42	18	22	21
Paper	55	42	46	25	25	31
Non-woven	55	69	78	21	25	22
Woven	73	75	80	27	23	20

Table 5.2 Disposal options opined by consumers in China, HK and India

Options in %	Plastic			Paper			Nonwoven			Woven		
	CH	HK	IN	CH	HK	IN	CH	HK	IN	CH	HK	IN
Recycle	24	21	33	46	45	47	33	27	29	32	27	26
Reuse	32	30	31	32	38	22	49	54	44	53	53	53
Landfill	44	49	36	22	17	31	18	19	27	15	20	21

more than 90 %, of the respondents in all three countries would like to support recycling and are wiling to place the used bags in recycling bins. Here comes the support of governmental policies to facilitate recycling options in each country by means of encouraging recycling, providing more recycling bins and so on. From the survey, it was noticed that in India and China, around 50 % users only confirmed the provision of recycling systems and in HK only 66 % confirmed. This is one of the major outcomes from this survey, from which further actions should be taken by respective governments to promote recycling in their countries.

5.4 Consumption Behaviour and Eco-Impact: Modelling of Life Cycle Impacts

Consumer behaviour plays a significant role in influencing eco-impacts of any product and shopping bags is not an exception. Degree of influence of life cycle impacts of grocery shopping bags by consumer behaviour would be more, since they possess a short span of life. Importance of consumer behaviour in deciding different environmental impacts made by shopping bags was studied by a systematical study conducted by authors.

This study was conducted to study the influence of consumer behaviour on eco-impacts of various shopping bags used for grocery purposes under three different options namely [15, 16]:

- Usage and disposal criteria with the existing usage behaviour to reuse and governmental policies to recycle (option 1),
- Usage and disposal criteria from the consumer's perspective and with recycling systems in place (option 2),
- Usage and disposal criteria in the absence of recycling systems (option 3).

This study was conducted for both single use bags and reusable bags. Plastic and paper bags in single use variety and non-woven polypropylene and woven cotton bags in reusable category were evaluated in this study under two different scenarios. The first scenario, which was considered as base line scenario, involved the study of the eco-impact of these bags in their manufacturing phase alone, without considering their usage and disposal phases and the second scenario revolved around the cradle to grave stages including use and disposal phases with the three options mentioned above.

Life cycle assessment study was conducted in full scale for this attempt. Cradle to gate stage data were referred from secondary data sources, i.e. from the final report prepared for Environment Australia in 2002 and its updated version prepared in 2004 [3, 4]. Data pertaining to cradle to gate stage focuses on main issues such as material consumption, energy needed for manufacturing process, green house gas emissions. Use and disposal values for this assessment were obtained from the consumer survey and the eco-impacts were evaluated in three different

Table 5.3 Life cycle inventory data of grocery shopping bags (cradle to gate)

Type of bag	Unit weight	Number of bags per year	Material consumption	GHG (CO_2 emissions)	Primary energy required
Plastic bag	6 g	520	3,120 g	6.08 kg	210 MJ
Paper bag	42.6 g	520	22,150 g	11.8 kg	721 MJ
Non-woven bag	65.6 g	4.15	272.24 g	1.96 kg	46.3 MJ
Woven cotton bag	125.4 g	9.1	1,141.14 g	2.52 kg	160 MJ

options highlighted above. The scope and boundaries of this study included the LCI obtained from the available data collected from secondary data sources. The functional unit of this study was defined as the sufficient capacity for a household consuming approximately 70 grocery items which were carried away from a supermarket in shopping bags every week for 52 weeks [3, 4].

Data for life cycle inventory pertaining to cradle to gate stages of different grocery shopping bags, focused chiefly on the energy and pollutants details for the functional unit assumed was referred from the previous studies [3, 4] and listed in Table 5.3.

From the survey results, usage and disposal values were deduced and formulated for the three options selected for this study as shown in Tables 5.4, 5.5, 5.6.

Tables 5.3, 5.4, 5.5, 5.6 constituted the life cycle inventory of various grocery shopping bags and these inventory details were processed by 7.2 version of simapro software used for life cycle assessment calculations. Eco-indicator'99, a damage oriented method was employed to assess the environmental impacts made

Table 5.4 Values for usage and disposal options of grocery shopping bags according to existing possibilities [15, 16]

Percentage	Plastic bags			Paper bags			Non-woven bags			Woven bags		
	CN (%)	HK (%)	IN (%)	CN (%)	HK (%)	IN (%)	CN (%)	HK (%)	IN (%)	CN (%)	HK (%)	IN (%)
Recycle	21	22	18	31	25	25	22	25	21	20	23	27
Reuse	46	42	55	42	38	28	78	69	55	80	75	73
Sent to landfill	33	36	27	27	37	47	0	6	24	0	2	0

Table 5.5 Percentage of grocery shopping bags that can be recycled/reused/sent to landfill according to consumer's perceptions [15, 16]

Percentage	Plastic bags			Paper bags			Non-woven bags			Woven bags		
	CN (%)	HK (%)	IN (%)	CN (%)	HK (%)	IN (%)	CN (%)	HK (%)	IN (%)	CN (%)	HK (%)	IN (%)
Recycle	24	21	33	46	45	47	33	27	29	32	27	26
Reuse	32	30	31	32	38	22	49	54	44	53	53	53
Sent to landfill	44	49	36	22	17	31	18	19	27	15	20	21

Table 5.6 Values for usage and disposal options of grocery shopping bags in case of absence of recycling system [15, 16]

Percentage	Plastic bags			Paper bags			Non-woven bags			Woven bags		
	CN (%)	HK (%)	IN (%)	CN (%)	HK (%)	IN (%)	CN (%)	HK (%)	IN (%)	CN (%)	HK (%)	IN (%)
Recycle	0	0	0	0	0	0	0	0	0	0	0	0
Reuse	46	42	55	42	38	28	78	69	55	80	75	73
Sent to landfill	54	58	45	58	62	72	22	31	45	20	25	27

by different grocery shopping bags. Life cycle of these bags were modelled in five stages namely Characterisation, Damage Assessment, Normalisation, Weighing, Single Score Analysis by simapro software. From the cradle to gate assessment (which was considered as baseline scenario), plastic bags score out paper bags in single use bags category and non-woven bags surpassed woven cotton bags in reusable category as far as the reduced environmental impacts are concerned and the detailed results can be found in [15, 16]. Overall, non-woven bags made by polypropylene stand out amongst all alternatives considered for this study for the functional unit assumed in this study.

Different shopping bag alternatives were compared in terms of their eco-impacts assessed by Eco-indicator'99 method for the three options formulated above. Single score results presented in points (Pt) arising from the three options were compared with the baseline scenario for different bags. Table 5.7 lists the comparative results of the three options for different shopping bag alternatives.

From the detailed results of this study explained [15, 16], important points to be noted from this study is, if a bag is thrown or disposed without using or reusing it, its environmental concerns are huge. Reusing different bags to the maximum possible extent is the way to curb the environmental impacts made by them and more number of times if a bag is reused, it is even to compensate the eco-impacts created by the same bag in cradle to gate stage. Countries which opted more to reuse a bag created lower amount of eco-impacts tabulated in Table 5.7. One can notice the difference in all territories between options 1 and 2, which enumerates the existence of difference between actual scenario and consumer's perception. One of the best ways to interpret this assessment is through the three different options formed and to show the importance of recycling is to compare the results of options 1 and 3, which obviates the need to go for recycling. Results of option 3 are on higher side, due to the diversion of bags to landfill due to the absence of recycling systems. Governments play a major role in curtailing the eco-impacts made by various shopping bags in terms of promoting recycling policies and providing more recycling bins.

Table 5.7 Comparative results of eco-impact of various shopping bags [15, 16]

	Base line study	With usage and disposal options (options 1, 2 and 3) in China, Hong Kong and India								
Total points	Without usage and disposal criteria	China-option 1	China-option 2	China-option 3	HK-option 1	HK-option 2	HK-option 3	India-option 1	India-option 2	India-option 3
Plastic bags (Pt)	36	19.3	24.3	19.5	20.7	25	21	16	24.5	16.3
Paper bags (Pt)	126	72.9	85.5	73	78	77.9	78.1	90.6	98.1	90.7
Non-woven (Pt)	0.77	0.151	0.366	0.171	0.218	0.333	0.241	0.331	0.409	0.35
Woven (Pt)	2.48	0.497	1.17	0.505	0.622	1.18	0.632	0.67	1.18	0.682

5.4.1 Carbon Footprint of all Types of Grocery Shopping Bags

With the same assessment set up, authors performed carbon footprint assessment of various grocery shopping bags in China, Hong Kong and India. Same data sources were referred for cradle to gate stage and for use and disposal values were obtained from the consumer survey conducted in China, Hong Kong and India. Functional unit of this study was derived from the analysis of consumption statistics of shopping bags in these three territories. The following functional unit was earmarked for this assessment: "number of shopping bags used for grocery shopping per year by an average Chinese/Indian/HK resident" [17]. Based on the consumption statistics and further analysis made on equating different shopping bags to this functional unit, it was confirmed that 1,095 plastic and paper bags, 10.95 non-woven and 21.9 woven bags were required to fulfil the functional unit for an average Chinese and HK residents. For the same functional unit, Indians need 150 plastic and paper bags, 1.5 non-woven and 3 woven bags [17].

Cradle to gate results were recalculated from the secondary data sources [3, 4] to suit this functional unit assumed for China, Hong Kong and India. Values from the consumers according to the existing possibilities, listed in Table 5.4 were utilised to complete the details of use and disposal phases. LCI was processed by Simapro 7.2 version of LCA software to calculate the carbon footprint values of 20, 100 and 500 years. Results of baseline study (without usage and disposal values) and with consumer's values on usage and disposal phases are listed in Table 5.8. For better clarity and to simplify, only the results of 100 years are presented in Table 5.8.

The conclusions listed above for eco-impact study are applicable here as well. Though it is well known that more number of reuse less will be the eco-impact or carbon footprint, however conducting a study to reveal that how much or to what extent reuse/recycle can benefit in reducing the environmental impacts, was the motive of the authors to perform these studies.

Table 5.8 Carbon footprint results of different shopping bags

Type of bag	GWP in 100 years (KgCo2 Eq) for China, HK			GWP in 100 years (KgCo2 Eq) for India	
	Baseline	With use and disposal		Baseline	With use and disposal
		CN	HK		
Plastic bags	523	282	303	72	32
Paper bags	1,744	1,030	1,100	239	176
Non-woven bags	145	32	45	20	9
Woven bags	448	90	112	61	17

Table 5.9 Carbon footprint results before and after plastic bags ban in China [20]

	Before ban	After ban
Usage of plastic bags	1,095	365
Kg CO_2 eq. for 100 years	521	174

5.5 Before and After Plastic Bags Ban in Various Countries: Results of Life Cycle Assessment Modelling

Consumer behaviour is governed by many factors, most of them come from the attitude of consumers. Apart from this, governmental regulations and laws demand consumers to change their attitude and behaviour towards certain products. One such case is the influence of plastic bags ban on consumer behaviour. All over the world almost, governments of all countries have started introducing plastic bags ban in terms of levying some amount of money to restrict the usage of plastic bags and most of the countries have completely banned the use of plastic bags. It was noticed that after the introduction of ban or levi, the usage of plastic bags have gone down significantly [18].

Authors performed many studies on this subject and just to show case an example, a case study performed in China before and after plastic bags ban is discussed here. It was learnt from a reference that exactly after one year of implementation of ban in China, the number of plastic bags usage has dropped to 1/3 level [19]. Based on the consumption statistics in China, it was noticed that an average consumer uses 1,095 plastic bags/year. After the ban, this number went down to 365 bags/year [20]. Calculations of LCA were repeated to accommodate this change and the carbon footprint was modelled with the aid of 7.3 version of Simapro software and the results are tabulated in Table 5.9 [20]. From the results, one can understand the result of change in consumer behaviour due to governmental influence and its corresponding effect on carbon footprint.

5.6 Concluding Remarks

Consumer behaviour influences the life cycle impacts of consumer products to a greater extent. It is worthwhile and meaningful to conduct life cycle assessments using the values obtained from the actual users to model the use and disposal scenarios of products. It is far better than performing a study with assumptions, which will no way reflect the true behaviour and cannot be used as a benchmark to change anything. Aspects related to the influence of consumer behaviour on eco-impacts of grocery shopping bags were discussed in detail in this chapter. A consumer survey conducted in China, Hong Kong and India amongst different user groups of shopping bags and its detailed results were discussed in this chapter. This chapter also highlighted the influence of consumer behaviour on eco-impacts and carbon footprints of various grocery shopping bags with the aid of life cycle assessment. Influence of plastic bag ban on consumer behaviour and the corresponding change in environmental impacts was also discussed briefly in this chapter.

References

1. Chaffee C, Yaros BR (2007) Life cycle assessment for three types of grocery bags—recyclable plastic; compostable, biodegradable plastic; and recycled, recyclable paper. Final report, Boustead Consulting and Associates Ltd
2. Ecobilan (2008) Evaluation study of environmental impacts of paper and plastic carrier bags. Available at. http://www.eurosac.org/eurosac/pdf/3808_LCA-Carrier-Bags-Methodology-(english).pdf
3. Nolan ITU, centre for design (RMIT), and Eunomia Research and Consulting Ltd (2002) Plastic shopping bags—analysis of levies and environmental impacts. Environmental Australia, Department of the Environment and Heritage, Commonwealth Government: Canberra
4. ExcelPlas Australia, centre for design (RMIT), and Nolan ITU (2004) The impacts of degradable plastic bags in Australia, final report to Department of the Environment and Heritage, Commonwealth Government of Australia: Canberra
5. James K, Grant T (2005) LCA of degradable plastic bags, 4th Australian LCA conference, Sydney, Feb 2005
6. FRIDGE (2002) Socio-economic impact assessment of the proposed plastic bag regulations report for the national economic development and labour council. Bentley west management consultants, Johannesburg, South Africa. Available at. sfenvironment.org/downloads/library/asticlifecycleanalysis.pdf
7. GUA (2005) Gesellschaft für umfassende Analysen the contribution of plastic products to resource efficiency, Vienna. Available at. http://www.plasticseurope.org/Content/Default.asp?PageID=517
8. Los Angeles County (2007) An overview of carryout bags in Los Angeles County, A Staff Report to the Los Angeles County Board of Supervisors, Available at. ladpw.org/epd/pdf/PlasticBagReport.pdf
9. Paper vs. Plastic bags (2004) downloaded at. http://iere.org/ILEA/lcas/franklin1990.html
10. The ULS Report (2008) Review of life cycle data relating to disposable, compostable, biodegradable and reusable grocery bags, 28 Mar 2008
11. Ellis S, Kantner S, Saab A, Watson M (2005) Plastic grocery bags: the ecological footprint, student publications, VIPIRG publications, University of Victoria, PO Box 3050 STN CSC, Victoria
12. Dilli R (2007) Comparison of existing life cycle analysis of shopping bag alternatives, final report, Hyder Consulting Pty Ltd, Australia, April 2007
13. Muthu SS, Li Y, Hu JY, Mok PY (2010) An exploratory comparative study on eco-impact of paper and plastic bags. J Fibre Bioeng Inform 4:307–320
14. Li Y, Muthu SS, Hu JY, Mok PY, Ding X, Wang X, Weibang C (2010) Eco impact of shopping bags: consumer attitude and government policies. J Sustain Dev 3:71–83
15. Muthu SS, Li Y, Hu JY, Mok PY, Ding X (2012) Eco-impact of plastic and paper bags. J Eng Fibres Fabr 7(1–2012):26–37
16. Muthu SS, Li Y, Hu JY, Mok PY, Ding X (2012) Influence of consumer behaviour and governmental policies in China, Hong Kong and India: an eco-impact study of reusable (nonwoven and woven) shopping bags. Energy Educ Sci Technol Part A Energy Sci Res 28(2):1131–1144

17. Muthu SS, Li Y, Hu JY, Mok PY (2011) Carbon footprint of shopping (grocery) bags in China, Hong Kong and India. Atmos Env 45(2011):469–475
18. Plastic bag reduction around the world, downloaded at. http://www.marrickville.nsw.gov.au/marrickville/internet/resources/documents/pdfs/bagbusters/around-the-world.pdf
19. West L (2009) China plastic bag ban shows strong results after first year, 2009. Available at. http://environment.about.com/b/2009/06/01/china-plastic-bag-ban-shows-strong-results-after-first-year.htm. 1 June 2009. Accessed 11 Mar 2010
20. Muthu SS, Li Y, Ze L (2013) Role of human factors in environmental sustainability: a case study of shopping bags consumption. Energy Educ Sci Technol Part B: Soc Educ Stud 5(2):793–808

Chapter 6
End-of-Life Assessment of Grocery Shopping Bags

6.1 Introduction

Management of solid waste is one of the nightmares faced by environmental scientists and governments. The issue of solid waste is aggravated by products which are regarded as a symbol of throwaway society such as shopping bags. Disposal of shopping bags is paid much attention by different parties such as public, government and environmentalists, since shopping bags made out of plastic materials do not biodegrade and they take a very long period to decompose. Such non-degradable materials pose severe threat to the environment, people and animals. Apart from the issue of biodegradability, limited landfill space on earth made almost all countries to perceive disposable or single use shopping bags as a potential threat to the environment. Because of these two main reasons, disposal or end-of-life stage of shopping bags is a hot topic and the need of the hour. It is one of the important elements in the agenda of any government to seriously consider the options to get rid-off the environmental threats posed by disposable or single use shopping bags.

When a shopping bag is decided to be disposed, it has many destinations to reach. The issue is to select the best destination which is beneficial to the environment primarily and secondly to benefit the customer economically. It should satisfy these dual needs. Available destinations at the end-of-life for any product on earth, including shopping bags are: reuse, recycle, incineration and landfill. The options are listed in the hierarchy of most beneficial to the economy and the environment. These options and other associated details pertaining to each impact in the light of grocery shopping bags will be discussed in this chapter.

6.2 Reuse and Impacts

Environmental assessments need not focus on or project only detrimental impacts, rather they deal with positive impacts as well. Positive impacts are known as benefits (Benefit to the environment). One of the elements involved in the life

S. S. Muthu and Y. Li, *Assessment of Environmental Impact by Grocery Shopping Bags*, EcoProduction, DOI: 10.1007/978-981-4560-20-7_6,

cycle assessment with positive impacts (benefits) is reuse. Even though recycling and incineration also bring benefits, they demand certain amount of resources to bring those benefits. But reuse option in the end-of-life does not demand any resources and brings ample benefits to the environment in terms of delaying the disposal of a product, delaying the arrival of a new material to be extracted, manufactured and processed and its associated impacts are prevented as well. One of the main advantages is if a product is reused instead of being disposed of, it extends the length of life of the product, which equates and compensates the actual life cycle impacts incurred in manufacturing the product with its prolonged life.

There are two types of reuse- primary and secondary. Primary reuse is applicable to the products which are reused for the same purpose. Secondary reuse means, if a product is discarded, if it will be used for some other than its the original function, for instance, worn out plastic bags will be used as dust bin liners. Reusable bags were invented for the purpose of eliminating or reducing the environmental impacts of single use disposable shopping bags. Reusable bags made out of non-woven technology with polyester and polypropylene can be reused for many times without or with a very little maintenance. Woven shopping bags made out of cotton, hemp can also be reused for many times, again without or with a few wash and care.

It has to be understood that the environmental concerns will be very huge if the reusable bags are disposed after single use or without using it for sufficient number of times, till the functionality limit of the bag permits. So, it would be very much beneficial if shopping bags are reused till they reach their maximum limit of usage. Afterwards, they must be considered for secondary reuses and after which they must be forwarded to recycling options, rather than disposing at landfills.

6.3 Recycle and Impacts

The next best option in terms of disposal of shopping bags is recycling. Recycling refers to the conversion of discarded products to usable ones. According to the definition by EPA of the U.S., recycling is the process of collecting and processing materials that would otherwise be thrown away as trash and turning them into new products [1]. Recycling brings ample benefits to the environment such as saving energy, conserves various resources, and reduces the amount of solid waste sent to landfill and incinerators. Recycling also reduces the amount of green house gas emissions and other environmental impacts and lead to sustainable development [2].

Recycling converts the materials or products at their end-of-life stage of their life cycle and transform them either into the same product or primary product (called as closed loop recycling) or as a secondary product (called as open loop recycling). It is one of the challenging tasks in recycling to produce a material from recycling, which is at par with the quality of the original or virgin raw material. Many factors decide the fate of a recycled material and in most cases, recycling is a down-cycling process. However, if a material is recycled into the same product or a different product, it intends to replace some designated or specific quantity of the virgin material and prevents the environmental impacts pertaining to the production of it, which is environmentally beneficial. Apart from this, recycling process prevents the material at the end-of-life being wasted and disposed onto landfills, which are liable to create many environmental impacts.

They are two types of waste which can be recycled and they are PIW (post industrial waste) and PCW (post consumer waste). Mostly retaining or achieving the quality of recycled materials at par with the virgin or original raw materials is relatively easy with PIW rather than with PCW, since in case of PCW, the quality of the material will be degraded, since they are the ones discarded after their useful life.

All types of grocery shopping bags can be recycled and they necessarily have to be recycled for the sake of environmental preservation. Recycling of plastic bags is one of the most popular topics being discussed widely in internet forums and blogs. However, still no research publication is available yet in the literature dealt about the facts and figures of recycling of different types of grocery bags along with the pros and cons of recycling of each type of bag.

6.4 Incineration and Landfill Impacts

Other destinations available at the end-of-life of shopping bags are incineration and landfill. Incineration involves waste destruction by burning in a furnace at high temperatures [3]. Incineration with energy recovery is the next feasible option of waste disposal at the end-of-life. However, incineration is not an well accepted method in many countries due to the environmental impacts (due to the release of carbon monoxide, dioxins and many other harmful substances due to incomplete combustion of waste [3]) arise from incinerators.

The last option with loaded environmental impacts and a nightmare to environmentalists, governments and even to public is landfill. Disposal to landfill is an undesirable option and prone to create umpteen number of issues to our living planet, hence must not be encouraged for waste disposal.

6.5 Biodegradation of Shopping Bags

Once it is decided to dispose of any product after its potential life span, they are expected and they must biodegrade smoothly to end their life. Though it is a common phenomenon expected from any product in earth, the severity of issue is high for products such as shopping bags, which represent the symbol of the throw-away society. Irrespective of the type of bag and its material content, a grocery bag necessarily has to biodegrade at the end of its life cycle. Hence it is pivotal to evaluate the capability of bags' to biodegrade. Further, it is essential to evaluate different grocery bags in terms of their biodegradation under the same platform, thereby they can be evaluated on a same scale and compared. The authors did attempt to evaluate the biodegradation of various shopping bags using the same platform [4] and the details of this study are briefly discussed here.

6.5.1 Soil Burial Test of Grocery Shopping Bags

Amongst various methods of measuring biodegradation, soil burial test is one of effective and cheap methods to measure the extent of biodegradation gone by different products. Many studies have utilized this test to assess the ability of the samples to biodegrade [5–12]. As per AACTCC 30 standard [13], an experimental setup was made in one of the labs of The Hong Kong Polytechnic University and the test was conducted in lab conditions for different shopping bag samples [4]. According to AATCC 30 standard, soil was procured from the market and its pH level was between 5 and 6.5 and electrical conductivity between 0.8 and 1.5. The soil was first dried before the test and during the whole period of testing, the moisture content of soil was maintained at 25 ± 5 % to comply with the conditions stipulated in the standard.

A soil bed was prepared to 13 cm depth in compliance with all the conditions stated in the standard to carry out this test. For this test, paper bag sample of 158.73 gm/m^2 (GSM) and woven cotton sample of 368.3 GSM, HDPE sample of 83.5 GSM, LDPE sample of 95.17 GSM, polypropylene nonwoven bag sample of 104.57 GSM and Nonwoven Polyester bag sample of 109.93 GSM were chosen. These chosen samples were cut into 2.5 × 15 cm pieces and 4 samples from each category of shopping bag were prepared. Samples were buried in the soil at 3 cm depth [10 cm depth and top layer must be of 3 cm soil] and allowed to degrade for periods of 0–90 days. To ensure the sufficient moisture in soil during test, moisture content was checked at regular intervals throughout the entire test period with the aid of a moisture meter. During the entire test period of 90 days, the temperature of the setup was maintained at temperature: 28 ± 2 °C with the help of a temperature control made for the test chamber [4].

After the entire designated degradation period, samples were rinsed with distilled water and dried. All the samples were tested for weight loss after 0, 30, 60 and 90 days and loss/gain in tensile strength, strain and extension for 0 and after 90 days. Changes in the appearance were measured with the aid of a Microscope and the extent of biodegradation was evaluated in terms of both loss in weight and tensile strength. Tensile strength of the samples before and after the biodegradation of 90 days was measured as per ASTM D 5034[Grab Test] (ASTM D5034–09).

Results of tensile strength measurement and weight measurement are described in Figs. 6.1, 6.2, 6.3, 6.4 along with % loss in weight and strength after 90- days for different samples considered for this study. From the results, it is noticed that paper biodegraded completely. It was observed during the experimentation that after just a week, paper samples started disintegrating and at the end of the test

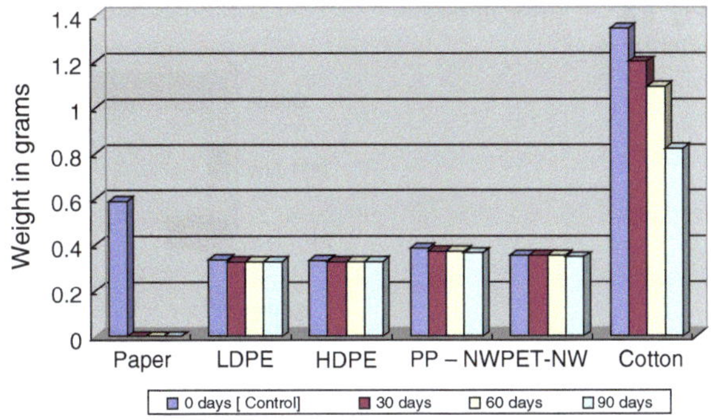

Fig. 6.1 Weight loss result of different grocery bags

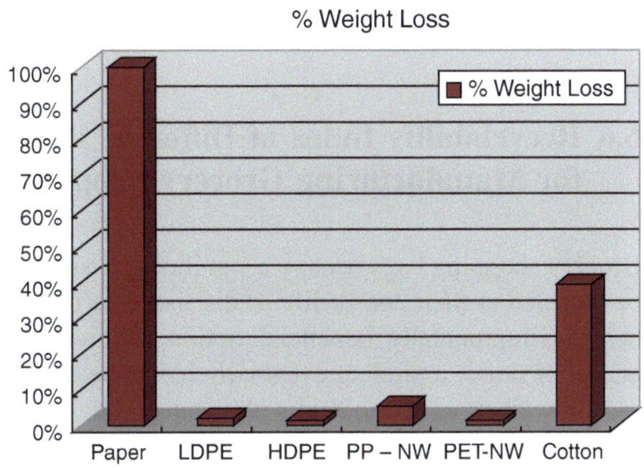

Fig. 6.2 % weight loss after 90 days

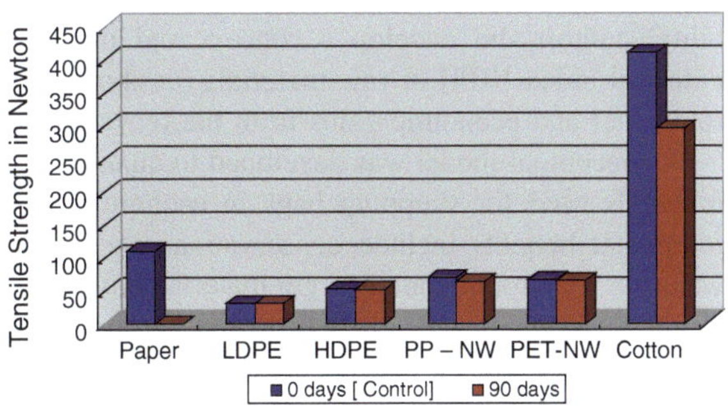

Fig. 6.3 Tensile Strength result of different grocery bags

period, they were completely disintegrated. Cotton samples also showed significant weight and strength loss, which is a positive indication of biodegradation. Polypropylene appeared to be better in terms of biodegradation in the synthetic materials category with 9 % strength loss recorded after 90- days.

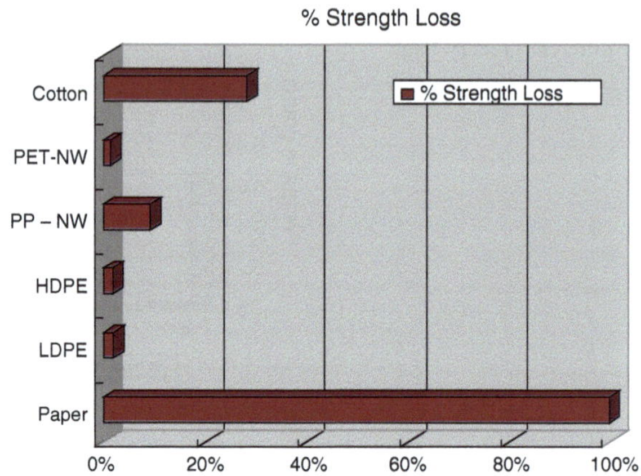

Fig. 6.4 % Strength loss after 90 days

6.6 Recyclability Index of Different Raw Materials used for Manufacturing Grocery Shopping Bags

Grocery shopping bags occupy a significant position in daily activities and need to be recycled at their end-of-life as discussed earlier. Recycling is the next possible best environmentally beneficial option followed by reuse. Even, after certain period of reuse, a product eventually has to be disposed, in that case, recycling is certainly a best option to deal with the disposal. Many types of fibers/raw materials are being used to manufacture shopping bags and their potential recyclability of varies from one material to another and many factors govern their recyclability. Since no models or studies are available in the literature to quantify the recyclability, authors did develop a concept and model to quantify the recyclability potential index [RPI] of raw materials for shopping bags, considering their environmental and economic gains from the recycling process [13].

A conceptual model was developed to quantify the recyclability of various raw materials used for shopping bags in economic and environmental terms. Environmental benefits include conservation of essential resources such as energy, water for the production of virgin material, environmental impact of production of virgin material include ecological and carbon footprint, human health impacts, environmental impact of land filling the material instead of recycling and finally the benefit of recycled materials in terms of energy conservation compared to virgin materials were considered to develop the model. The monetary benefits of recycled materials proposed by Villaba et al. [14] were considered for calculating the economic benefits of recycled materials. In this study, the proposed concept of a Recyclability Potential Index [RPI] was tested for ten important textile fibres such as nylon 6 and 66, viscose, acrylic, polyester, wool, cotton, polypropylene, polyethylene's [LDPE and HDPE] and paper used for manufacturing shopping bags.

6.6.1 Basic Concept of Recyclability Potential Index

A Recyclability Potential Index [RPI] is a composite factor, taking into account numerous factors from various perspectives and cannot be defined with one or two parameters. Though there are many possible factors to be looked at, only environmental and economic sides were taken into consideration to derive RPI [13]. As defined by authors [13], RPI can be calculated as follows:

$$RPI = \sum EGI_1 + EGI_2. \tag{6.1}$$

where EGI_1—Environmental Gain Index and EGI_2—Economic Gain Index.

Further,

$EG_1 =$	$\sum X_1 + X_2 + X_3 + X_4$
where $X_1 =$	Saving potential resources
$X_2 =$	Environmental impact caused by producing virgin fibres/materials
$X_3 =$	Environmental impact due to land filling
$X_4 =$	Environmental benefits gained out of recycling versus incineration
And, $EG_2 =$	x_1/x_2,

where x_1 = Price of recycled fibre/material; x_2 = Price of virgin fibre/material.

6.6.2 Quantification of Recyclability Potential Index of Raw Materials Used for Shopping Bags

As discussed earlier, various data were collected for environmental and economical gain indices. In environmental gain index, first, water and energy needed to produce 1 kg of different raw materials were collected. Followed by this, environmental impacts of producing virgin materials (to produce 1 kg of virgin material) were quantified for three indicators namely ecological, carbon footprints and ecological damage in terms of human health with the aid of Simapro 7.2 software. Ecological footprint was modeled by Ecological Footprint V1.00, carbon footprint was modeled by IPCC 2007 GWP 100a method and ecological damage was quantified by the Ecoindicator'99 method, where only human health impacts were considered.

The third factor was to stress the importance of recycling by means of quantifying the impacts of landfilling. For this, ecological, carbon footprints and ecological damage in terms of human health of disposing 1 kg of selected raw materials at landfills were modelled by Simapro 7.2 version of LCA software. The final factor considered was the environmental benefit of recycling versus incineration. Different raw materials selected were assessed in terms of environmental benefits of Recycling Vs Incineration with the aid of literature data.

For economic gain index, as stated earlier, values of both recycled and virgin materials obtained from the market were obtained for all the raw materials under

Energy [MJ]		Water [kgs.]		E.I. of Virgin – EFP	
<50	1	<100	1	<5	1
51- 100	2	101-200	2	5.1-10	2
101-150	3	201-300	3	10.1-20	3
151-200	4	301-400	4	20.1-30	4
>201	5	>401	5	>30.1	5
E.I. of Virgin – CFP		E.I. of Virgin – HHI		E.I. of Landfill-EFP	
<2	1	<20	1	<50	1
2.1-4	2	21- 40	2	51- 100	2
4.1-6	3	41-60	3	101-150	3
6.1-8	4	61-80	4	151-200	4
>8.1	5	>81	5	>201	5
E.I. of Landfill – CFP		E.I. of Landfill– HHI		Energy Conserved	
<100	1	<20	1	>15001	1
101- 300	2	21- 40	2	15000-11001	2
301- 500	3	41-60	3	11000-7001	3
501 -700		61-80	4	7000-3001	4
4 >701		>81	5	<3000	5
5					

EGI_2	
>0.81	1
0.8-0.61	2
0.6-0.41	3
0.4-0.21	4
<0.20	5

Fig. 6.5 Scaling template developed for RPI model, (E.I—Environmental impact; CFP—Carbon Footprint; EFP—Ecological Footprint; HHI—Human Health Impacts)

consideration for this study. With all these data for both indices, many values are available with different units under different headings. To unify them and to obtain separate indices of EGI_1 and EGI_2 and consequently obtain RPI as defined earlier, a scaling template was developed (Fig. 6.5). This scaling template consists of five scales ranging from 1 to 5 [13].

With the aid of the scaling template created and the equations of RPI developed, different raw materials were quantified in terms of their environmental and economic gains and eventually their RPIs were also quantified. Interpretation of RPI follows the following rule: *lower the RPI value, the better the recyclability.* Table 6.1 gives the RPI values and the ranking in terms of recyclability of the chosen raw materials under question.

From Table 6.1, it is noticed that polyester and polypropylene score out all the different raw materials under consideration. They surpassed all the chosen raw materials in both the environmental and economic aspects considered in this research. Polypropylene scored ahead of polyester in environmental considerations and polyester outscored polypropylene in the economic aspect; thereby both jointly occupy best position in terms of RPI.

This model took into consideration of different factors for calculating environmental gain index, namely saving potential resources (water and energy),

Table 6.1 RPI of different raw materials [13, 14]

Raw material	Environmental gain index	Economic gain index	RPI	Ranking in terms of recyclability
Nylon 6	30	2	32	7
Nylon 66	33	4	37	9
Viscose	29	4	33	8
Acrylic	24	3	27	5
Polyester	20	1	21	1
Cotton	25	4	29	6
Wool	27	5	32	7
PP	19	2	21	1
LDPE	21	2	23	3
HDPE	20	2	22	2
Paper	20	4	24	4

various environmental impacts caused by producing virgin raw materials, environmental impacts due to disposal at landfills and environmental benefits gained out of recycling versus incineration. The model also considered another important index, economic gain index, which focuses primarily on the monetary value of recycled material/fibre vis-à-vis virgin material.

Though this model considered various important aspects needed to quantify the recyclability, still can be further improved by considering various other factors for calculating environmental gain index. Also, this model can be evaluated with different weighing factors for environmental and economic gains.

6.7 Concluding Remarks

This chapter dealt with the end-of-life options of grocery bags by discussing the various options at the end-of-life of grocery shopping bags. This chapter briefly discussed about the different end-of-life scenarios in the light of benefits and impacts to the environment from each scenario. Biodegradation study conducted by the authors to evaluate the extent of biodegradation of different grocery shopping bags was also discussed in this chapter. From the biodegradability assessment, it was noticed that the paper bags followed by cotton bags appear to biodegrade better after 90 days. This chapter also dealt with recyclability potential index concept defined by the authors and the model of RPI to evaluate the potential recyclability of different raw materials being employed to manufacture various grocery shopping bags. From this model, PP and PET were found to be bestowed with higher recyclability potential in terms of environmental and economic gains.

References

1. EPA (2008). Terms of environment: glossary, abbreviations and acronyms. Washington, DC: U.S. Environmental Protection Agency. Retrieved from: http://www.epa.gov/OCEPAterms/
2. EPA, recycling basics. Available at: http://www2.epa.gov/recycle/recycling-basics
3. Incineration, Dictionary definition. Available at: http://www.businessdictionary.com/definition/incineration.html
4. Muthu SS, Li Y, Hu JY, Mok PY (2012) A societal Hot-button issue: biodegradation (soil burial test) studies of grocery shopping bags. Energy Edu Sci Technol 29(1):31–40
5. Park CH, Kang YK, Im SS (2004) Biodegradability of cellulose fabrics. J Appl Polymer Sci 94:248–253
6. Azahari NA, Othman N, Ismail H (2009) Effect of blend ratio on physical properties and biodegradability of PVA/Corn starch composite. Malaysia polymer international conference (MPIC 2009), pp 506–512
7. Chen HL, Cluver B (2010) Microbiological resistance characteristics of naturally colored cottons. Textile Res J 80:1–7
8. Ismail H, Majid RA, Taib RM (2011) Effects of soil burial on properties of linear density polyethylene (LDPE)/thermoplastic sago starch (TPSS) blends. Pertanika J Sci Technol 19:189–197
9. Kim H-S, Yang H-S, Kim H-J (2005) Biodegradability and mechanical properties of agro-flour—filled polybutylene succinate biocomposites. J Appl Polymer Sci 97:1513–1521
10. Kumar R, Yakubu MK, Anandjiwala RD (2010) Biodegradation of flax fiber reinforced poly lactic acid. Express Polym Lett 4:423–430
11. Park CH, Kang YK, Im SS (2003) Effect of hydrophilicity on the biodegradability of polyesteramides. J Appl Polym Sci 90:2708–2714
12. Suh H, Duckett K, Bhat G (1996) Biodegradable and tensile properties of cotton/cellulose acetate nonwovens. Textile Res J 66:230–237
13. Muthu SS, Li Y, Hu J-Y, Mok P-Y (2012) Recyclability potential index (RPI): the concept and quantification of RPI for textile fibres. Ecological Indicators 18:58–62
14. Muthu SS (2011) Eco-functional Assessment of grocery shopping bags. PhD thesis, The Hong Kong Polytechnic University, Hong Kong

Chapter 7
Eco-Functional Assessment of Grocery Shopping Bags

7.1 Introduction

This chapter deals with the crux of this book, i.e. eco-functional assessment. As discussed earlier, eco-functional assessment includes the consideration of both ecological and functional properties on a single platform. This comprehensive assessment requires the details pertaining to entire life cycle phases of a product, as discussed in previous chapters from cradle to grave stages. This chapter includes the details of the concept of eco-functional assessment, framework developed for eco-functional assessment, modelling and simulation of eco-functional assessment. One of the main aims of this eco-functional research dealt by authors is to derive an eco-functional grade or score of any product and will be demonstrated by taking grocery shopping bags as an example here. This chapter also deals with the methodology for the derivation of eco-functional index or score and the interpretation of these eco-functional assessment results.

7.2 Concept of Eco-Functional Assessment

This study revolves around the concept of Eco-functional assessment, which is the central region formed by three interrelated aspects, i.e. functional, ecological properties and consumption behavior. The basic concept of eco-functional assessment, which is the interaction among these three aspects, is depicted in Figs. 7.1, 7.2. Interrelation and interaction of all these three aspects form the essence of eco-functional assessment as depicted in Fig. 7.1.

Many aspects need to be considered for functional and ecological aspects, which actually govern the consumer behavior. Assessment of all these three aspects form the crux of eco-functional assessment as indicated in Fig. 7.2.

S. S. Muthu and Y. Li, *Assessment of Environmental Impact by Grocery Shopping Bags*, EcoProduction, DOI: 10.1007/978-981-4560-20-7_7,

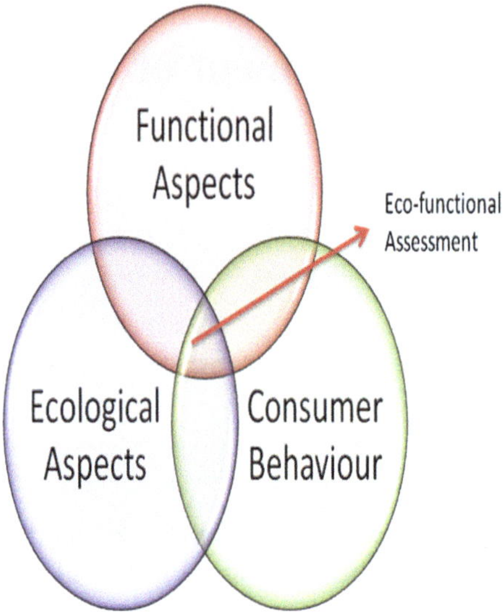

Fig. 7.1 Eco-functional assessment-interaction of aspects

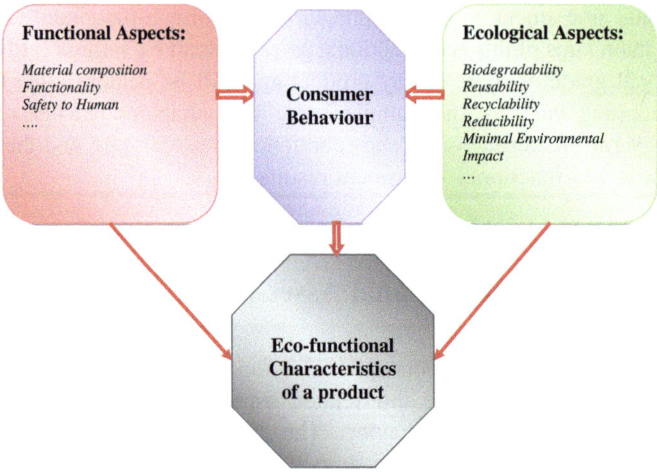

Fig. 7.2 Basic concept of eco-functional assessment

7.3 Development of Theoretical Framework of Eco-Functional Assessment

As discussed earlier, the concept of eco-functional assessment mainly deals with the consideration of functional, ecological parameters and consumer behavior to decide the environmental and human impacts made by a product in its life time.

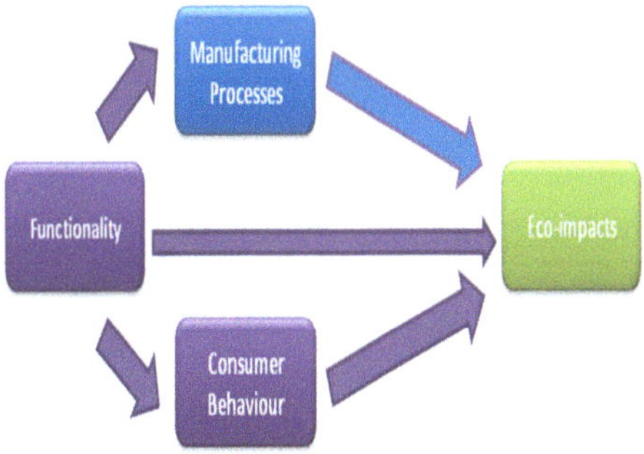

Fig. 7.3 Principle of eco-functionality assessment

This relies on the cradle to grave approach, which is one of the main variants of life cycle assessment. In cradle to grave approach, all the life cycle phases starting from raw material extraction, all the other phases namely manufacturing processes, transportation, use and disposal phases have to be considered for the assessment of impacts. In this assessment chain, the influence of manufacturing process on the environmental impact has been well established (Blue coloured in Fig. 7.3). However, the influence of functionality on manufacturing process and also the consumption behaviour (Violet coloured in Fig. 7.3), which in turn influences the ecological impact, has not been studied and reported in the literature so far and this was carried out by the authors.

The theoretical framework of Eco-functional assessment developed by authors is shown in Fig. 7.4. As shown in Fig. 7.4, the traditional life cycle assessment approach is already well established (shown in green colour). However, the functionality of the product assumes significant importance in each phase of life cycle and its influence on the entire life cycle necessarily needs to be evaluated. Hence, the eco-functional approach is of definite importance to study the environmental impacts of the products, which is discussed in this research work.

7.4 Development of Eco-Functional Model

The eco-functional assessment model was developed based on the theoretical framework depicted in Figs. 7.3, 7.4 to evaluate the eco-functional properties of different shopping bags considered for this research work and also to ascertain generally the eco-functional scores of those shopping bags. This model can be applied for any consumer product on earth and its applications are demonstrated for grocery shopping bags. The eco-functional assessment model developed comprises of four inputs and five outputs, as shown in Fig. 7.5. Raw materials, process of manufacture, functional properties and ecological properties are the four inputs considered for this model and quality, functionality, 3Rs, human impact and environmental impact, i.e. carbon footprint, ecological footprint and eco damage are the five outputs considered for this eco-functional model.

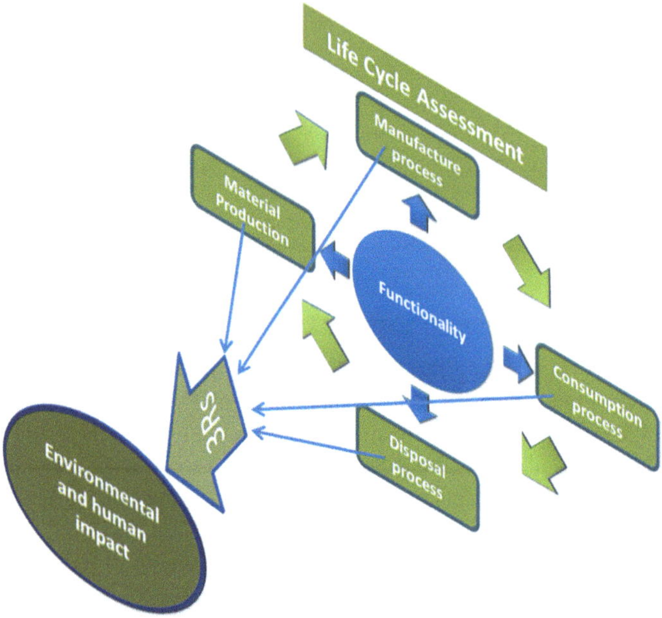

Fig. 7.4 Theoretical framework of eco-functional assessment [1]

Formulae, methods and rules were established for the eco-functional model framework in accordance with the relevant international standards and/or industry standards and the theoretical model of the product life cycle approach. According to the results calculated from the developed model, it is possible to determine the quality and functionality of products and also to assess their impact on human

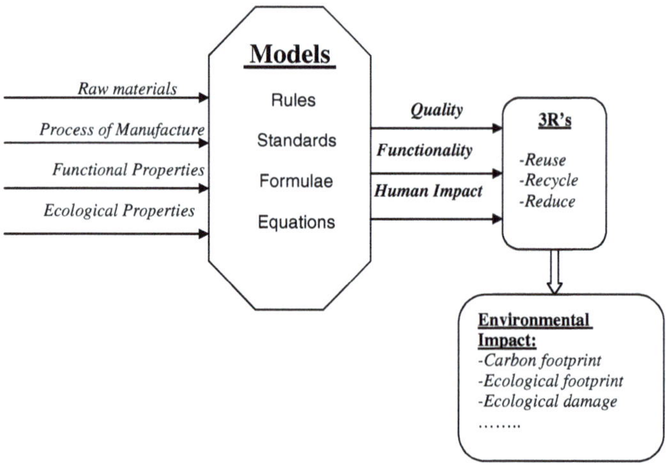

Fig. 7.5 Eco-functional model

Table 7.1 Ways of connection of inputs and outputs

Inputs of the model	Outputs to be connected	Ways of connecting inputs and outputs
Raw materials	1. Human impact 2. Environmental Impact	Both outputs are connected by a set of rules (logical rules)
Process of manufacture	1. 3Rs 2. Human impact 3. Environmental impact	1. 3Rs by rules (logical rules) 2. Human toxicity by formulae/equations 3. Environmental impact by formulae/ equations
Functional properties	1. Quality 2. Functionality	Both outputs are connected by rules (logical rules)
Ecological properties	1. 3Rs 2. Human impact 3. Environmental impact	Three are connected by a set of rules (logical rules)

beings and the environment. It is also possible to analyze the potential of products to comply with the 3Rs (Reuse, Recycle and Reduce). The toughest task of this model is to connect the inputs and outputs, the ways by which they are linked are shown in Table 7.1.

7.5 Modelling and Simulation of Eco-Functional Assessment

As discussed earlier, this model consists of four inputs and five outputs. It is the first step to develop the inputs and later on they need to be connected with the outputs earmarked for this assessment based on the rules framed in Table 7.1.

7.5.1 Values of Inputs

a. *Fibre/Raw material*

The first input is raw material used for the manufacture of the shopping bags. Values of EI and ESI of different raw materials were obtained from the model developed by the authors, discussed in Chap. 3.

b. *Process of manufacture*

Process of manufacture, which is being employed to manufacture the end products, is the second input of this model. To obtain the values of this input, the different production processes employed to manufacture different types of

Table 7.2 Functional properties of shopping bags [1]

Material composition	ISO 1833-1 (ISO 1833-1:2006)
Tensile strength and elongation of material	ASTM D 5034(Grab Test) (ASTM D5034–09)
Tear strength	ASTM D 1424 and 1922 (ASTM D1922–09; ASTM D1424–09)
Thickness	ASTM D1777 (ASTM D1777–96(2007)
Areal density (weight)	ASTM D3776 (ASTM D3776/D3776 M–09)
Bursting strength	ISO 13938-2 (ISO 13938-2:1999)
Colour fastness to friction/rubbing	ISO105 X12 (ISO 105-X12:2001)
Colour fastness to water	ISO 105 E 01 (ISO 105-E01:1994)
Colour fastness to washing	ISO 105 C 06 -B2 (ISO 105-C06:2010)
Colour fastness to perspiration results—acid and alkali	ISO105 E 04 (ISO 105-E04:1994)
Colour fastness to light	ISO 105 B02 (ISO 105-B02:1994)
Impact strength	From the eco-functional tester developed by authors
Load carrying capacity/weight-holding capacity	From the eco-functional tester developed by authors
pH	ISO 3071 (ISO 3071:2005)
Formaldehyde	ISO 14184-1 (ISO 14184-1:1998)
Waterproof	AATCC 127 (AATCC Test Method 127-2008)
Air permeability	ISO 9237 (ISO 9237:1995)
Water vapour permeability	ASTM E 96 (ASTM E96/E96M–10)

shopping bags (plastic, paper, nonwoven and woven bags) were studied in terms of accounting for their comprehensive life cycle inventory. Important areas that need to be considered at this juncture are amount of raw materials employed, quantities of energy water, additives and other materials consumed, amount of airborne wastes, solid, liquid and other wastes emitted, etc. Life cycle inventory of various shopping bags were collected and various values such as carbon footprint and ecological resources footprint were calculated by impact values pertaining to the average Chinese consumer for LCA calculations, discussed in Chap. 3.

c. *Functional properties*

The third input comprises the functional properties of shopping bags, which were obtained from the results of the tests, as shown in Table 7.2 [1]. Details of these test results were discussed in Chap. 4.

d. *Ecological properties*

The fourth input focuses on the ecological properties of shopping bags, which were taken from the results of the tests (discussed in Chap. 6) given in Table 7.3.

Table 7.3 Ecological properties of shopping bags [1]

Biodegradation of material	AATCC 30-soil burial test (AATCC 30:2004)
Reusability	From the eco-functional tester developed by authors
Recyclability	From the RPI model developed by authors

Table 7.4 Rules for input-1

Rule No.	IF R_{ESIR} is		R_{EI} and R_{HI} is
1	1	THEN	Close to none
2	2		Very less
3	3		Less
4	4		Moderately less
5	5		Moderate
6	6		Moderately high
7	7		High
8	8		Very high
9	9		Extreme
10	10		Extremely high

7.5.2 Input and Output Variables Linking

Each input needs to be connected with certain outputs in specific ways and the details are explained below. Table 7.4 shows the rules framed for the first input, i.e. fibre/other raw materials. Following the established rules, different raw materials were classified in terms of the ESIR and the outputs are environmental and human impacts (R_{EI} and R_{HI}).

For input no. 2, the process of manufacture, following are the the relevant outputs, which were developed for connection with second input: Human Impact in terms of Human Toxicity Potential; Environmental Impact (From LCA) in terms of Carbon footprint, Ecological footprint, Environmental burden—Emissions and Resources. The two outputs are connected by the equations well defined in LCA methodology [2]. Based on the calculation for Chinese consumers for these parameters, as defined in Chap. 3, values for these impacts were calculated and the rules stipulated in Table 7.5 were framed.

For the input no.3, Table 7.6 shows how input no.3 links to the relevant outputs, i.e. quality, functionality and human safety. Table 7.6 shows the range of values of different tests taken from Chap. 4 and the pass/fail criteria for various tests is shown in Table 7.7.

For the 4th input, Table 7.8 describes how input no.4 is linked to the relevant outputs, i.e. human toxicity, environmental impact and 3R's. Table 7.9 shows the range of values of different tests obtained from Chap. 6 and the pass/fail criteria for various tests.

Table 7.5 Range of values of LCA parameters and pass/fail criteria for input-2 [1]

Parameter	Range	Pass/fail criteria
Carbon footprint	118–9,650 g/CO_2	<50 % Pass >50 % Fail (R_{EI})
Ecological resources footprint	ADP Chinese—0.00117–0.1 g antimony eq./kg	<50 % Pass >50 % Fail (R_{EI})
Environmental load units	Emissions—0.0076–0.33 per kg	<50 % Pass >50 % Fail (R_{EI})
Human Toxicity	68.6–5,891.1 g 1,4-DCB eq./kg	<50 % Pass >50 % Fail (R_{HI})

Table 7.6 Linkage of outputs to input-3 [1]

Test	Criteria	Output
Material composition	PASS (meets the declaration)	Quality (R_Q)
Tensile strength and elongation of material	PASS (meets the requirement)	Functionality (R_F)
Tear strength	PASS (meets the requirement)	Functionality (R_F)
Thickness	PASS (meets the requirement)	Quality (R_Q)
Weight	PASS (meets the requirement)	Quality (R_Q)
Bursting strength	PASS (meets the requirement)	Functionality (R_F)
Colour fastness to friction/rubbing	PASS (meets the requirement)	Quality (R_Q)
Colour fastness to washing	PASS (meets the requirement)	Quality (R_Q)
Colour fastness to water	PASS (meets the requirement)	Quality (R_Q)
Colour fastness to perspiration	PASS (meets the requirement)	Quality (R_Q)
Impact resistance and toughness	PASS (meets the requirement)	Functionality (R_F)
Load carrying capacity	PASS (meets the requirement)	Functionality (R_F)
pH	PASS (meets the requirement)	Human Safety (R_{HI})
Formaldehyde	PASS (meets the requirement)	Human Safety (R_{HI})
Waterproof	PASS (meets the requirement)	Functionality (R_F)
Air permeability	PASS (meets the requirement)	Functionality (R_F)
Water vapour permeability	PASS (meets the requirement)	Functionality (R_F)

To arrive at the final results, three steps are required. The first step is to integrate quality and functionality to obtain the combined result (R_{QF}). The second step is to integrate human toxicity, environmental impact and the 3Rs to obtain the combined result (R_{EI}). The last step is to combine R_{QF} and R_{EI} to achieve $R_{Product}$, which is the ultimate, desired result from this developed eco-functional model. From the final result of $R_{Product}$, it is possible to determine the position of any shopping bag in terms of its eco-functionality [1]. Tables 7.10, 7.11, 7.12 list those three steps and Table 7.13 lists the decision criteria to be followed.

Table 7.7 Range of values of various tests and pass/fail criteria for input-3 [1]

Test	Range	Pass/fail Criteria
Material composition/fibre content	No range	If it meets the declaration (±5 %)-PASS (R_Q) Does not meet the declaration (±5 %)-FAIL
Tensile strength and elongation of material	1. Tensile strength- 60 N–766.1 N 2. Elongation-3.2–340.6 mm	<50 % Fail >50 % Pass (R_F)
Tear strength	0.5–45 N	<50 % Fail >50 % Pass (R_F)
Bursting strength	11.9–125 PSI	<50 % Fail >50 % Pass (R_F)
Impact strength	Absolute maximum capacity: 1–5 cycles; weight—2–5 kgs Comparative maximum capacity: 0–5 cycles; weight—3 kgs	<50 % Fail >50 % Pass (R_F)
Load bearing	4–35 kgs	<50 % Fail >50 % Pass (R_F)
Thickness	0.044–0.98 mm	<50 % Fail >50 % Pass (R_F) (R_Q)
Weight	39.7–368.3 g/m^2	If it meets the declaration (±10 %)-PASS (R_Q) (R_F) Does not meet the declaration(±10 %)-FAIL
Air permeability	0–789 mm/s	<50 % Fail >50 % Pass (R_F)
Water Vapour Permeability	14.3–1003.6 g/m^2.day	<50 % Fail >50 % Pass (R_F)
pH	5.92–9.12	4–9 Pass (R_{HI}) <4 and >9 Fail (GB 18401, 2010)
Formaldehyde	5.86–14.06 mg/kg	≤300 Pass(R_{HI}) >300 Fail (GB 18401, 2010)
Colour fastness to light	4	3 and >3 Pass(R_{HI}) (R_Q) <3 Fail
Colour fastness to friction	3–4—4–5 (3.5–4.5)	3 and >3 Pass(R_{HI}) (R_Q) <3 Fail (GB 18401)
Colour fastness to washing	4–4.5	3 and >3 Pass(R_{HI}) (R_Q) <3 Fail
Colour fastness to perspiration	Acid: 4–4.5 Alkaline: 4–4.5	3 and >3 Pass(R_{HI}) (R_Q) <3 Fail (GB 18401, 2010)
Colour fastness to water	4–4.5	3 and >3 Pass(R_{HI}) (R_Q) <3 Fail (GB 18401, 2010)

Table 7.8 Linkage of outputs to Input-4 [1]

Test	Criteria	Output
Biodegradation of material	PASS (meets the requirement)	Reduced human toxicity (R_{HI}) Lesser environmental impact (R_{EI})
Reusability	PASS (meets the requirement)	Reduced human toxicity (R_{HI}) Lesser environmental impact (R_{EI}) 3Rs (R_{3Rs})—Reusability
Recyclability	PASS (meets the requirement)	Reduced human toxicity (R_{HI}) Lesser environmental impact (R_{EI}) 3Rs (R_{3Rs})—Recyclability

Table 7.9 Range of values of various tests and pass/fail criteria for input-4 [1]

Test	Range	Pass/fail criteria
Biodegradation of material	Weight loss: 1.42–100 %	Weight loss: >50 % pass ($\mathbf{R_{HI}}$) ($\mathbf{R_{EI}}$)
	Tensile strength reduction: 1.36–100 %	<50 % Fail Tensile strength reduction: >50 % pass ($\mathbf{R_{HI}}$) ($\mathbf{R_{EI}}$) <50 % Fail
Reusability	Absolute maximum capacity: 3.7–100 cycles; weight- 4–35 kgs	<50 % Fail
	Comparative maximum capacity: 0–500 cycles; weight-10 kgs	>50 % Pass ($\mathbf{R_{3R's}}$) ($\mathbf{R_{EI}}$) ($\mathbf{R_{HI}}$)
Recyclability	Shown below (RPI rank-1–10)	1–5 PASS ($\mathbf{R_{3R's}}$) ($\mathbf{R_{EI}}$) ($\mathbf{R_{HI}}$) 6–10 FAIL

Table 7.10 Quality and functionality [1]

Rule No.	IF	Operand	R_Q/R_F		R_{QF}
1	R_Q is PASS	AND	R_F is PASS	THEN	GOOD
2	R_Q is PASS	AND	R_F is FAIL		POOR
3	R_F is PASS	AND	R_Q is FAIL		AVERAGE
4	R_Q is FAIL	AND	R_F is FAIL		POOR

Table 7.11 3Rs, Environmental impact, human impact [1]

Rule no.	IF	R_{EI}	R_{HI}		R_{EIF}
1	R_{3Rs} is PASS	R_{EI} is PASS	R_{HI} is PASS	THEN	GOOD
2	R_{3Rs} is FAIL	R_{EI} is FAIL	R_{HI} is FAIL		POOR
3	R_{3Rs} is PASS	R_{EI} is FAIL	R_{HI} is FAIL		POOR
4	R_{3Rs} is FAIL	R_{EI} is PASS	R_{HI} is FAIL		POOR
5	R_{3Rs} is FAIL	R_{EI} is FAIL	R_{HI} is PASS		POOR
6	R_{3Rs} is PASS	R_{EI} is PASS	R_{HI} is FAIL		AVERAGE
7	R_{3Rs} is FAIL	R_{EI} is PASS	R_{HI} is PASS		AVERAGE
8	R_{3Rs} is PASS	R_{EI} is FAIL	R_{HI} is PASS		AVERAGE

Table 7.12 Overall result [1]

Rule no.	IF	Operand	R_{EIF}/R_{QF}		$R_{Product}$
1	R_{QF} is GOOD	AND	R_{EIF} is GOOD	THEN	PASS
2	R_{QF} is GOOD	AND	R_{EIF} is POOR		FAIL
3	R_{QF} is AVERAGE	AND	R_{EIF} is POOR		FAIL
4	R_{QF} is AVERAGE	AND	R_{EIF} is AVERAGE		MEDIUM
5	R_{EIF} is AVERAGE	AND	R_{QF} is POOR		FAIL
6	R_{QF} is GOOD	AND	R_{EIF} is AVERAGE		PASS
7	R_{QF} is POOR	AND	R_{EIF} is GOOD		FAIL
8	R_{QF} is AVERAGE	AND	R_{EIF} is GOOD		PASS
9	R_{QF} is POOR	AND	R_{EIF} is POOR		FAIL

Table 7.13 Decision criteria [1]

Condition	Decision criteria (all of the conditions to be fulfilled)
R_Q is PASS	✔ Material composition/fibre content and GSM pass
R_F is PASS	✔ Tensile strength pass
	✔ Colour fastness to water pass
	✔ Colour fastness to perspiration (acid and alkali) pass
	✔ Colour fastness to washing pass
	✔ Colour fastness to friction pass
	✔Tear strength pass
	✔ Bursting strength pass
	✔ Impact strength pass
	✔ Water vapour permeability pass
R_{3Rs} is PASS	✔ Reusability pass
	✔ Recyclability pass
R_{EI} is PASS	✔ R_{EI} is close to none, very less, less, moderately less and moderate
	✔Biodegradation of material is pass
	✔Carbon footprint pass
	✔Ecological resources footprint pass
	✔ Environmental load unit-emissions pass
R_{HI} is PASS	✔R_{HI} is close to none, very less, less, moderately less and moderate
	✔ pH pass
	✔Formaldehyde pass
	✔ Biodegradation of material is pass
	✔Human toxicity pass

7.6 Derivation of Eco-Functional Index

Apart from the eco-functional assessment, with the aid of the developed eco-functional model, it is possible to derive an eco-functional index/score of any product in addition to evaluating the capacity of any product to sustain the requirements of the eco-functional assessment. The steps to arrive at the final eco-functional index are discussed below [1]. Different indices were defined for each inputs and they were detailed in Refs. [1, 2] to arrive at the final index called, "Eco-functional index".

The final result, eco-functional index is a result of the aggregation of the individual scores/indices of each input and is defined below:

$$\text{Eco-functional index} = \sum \text{ESI} + \text{HTI} + \text{EII} + \text{FI} + \text{ECI},$$

where

ESI Ecological sustainability index
EII Environmental impact index
HTI Human toxicity index
FI Functionality index
ECI Ecological index.

The grading systems and scales used for this index derivation can be found from the Refs. [1, 2].

7.7 Eco-Functional Assessment

Eco-functional assessment of various shopping bags under consideration was carried out with the aid of a computer programme, which was written using Microsoft Visual C++ 2008 Express version. An interface was created to connect the four inputs discussed previously with all the rules for assessment of the shopping bags in eco-functional terms and deriving the eco-functional index/score. The interface created is shown in Fig. 7.6 and the results of eco-functional assessment including index scores for all the grocery shopping bags considered in this research study are tabulated in Table 7.14. Codes used in this programming are given in Annexure 1.

7.8 Results and Interpretation of Eco-Functional Assessment

For this study, various types of grocery shopping bags of different weight categories, as stated in Chap. 4 were assessed for eco-functional assessment. From the results of eco-functional assessment, it was noticed that none of the bags considered for this study could meet the rules stipulated in this eco-functional model.

Detailed interpretation of results for each type of bag is discussed elsewhere [1, 2]. However, a brief note on paper bags' results is discussed here. As far as paper grocery shopping bags are considered, though they could not meet the requirements of functionality part, they scored very well in terms of environmental impact in terms of both raw materials and manufacturing process. They only fail to score in terms of

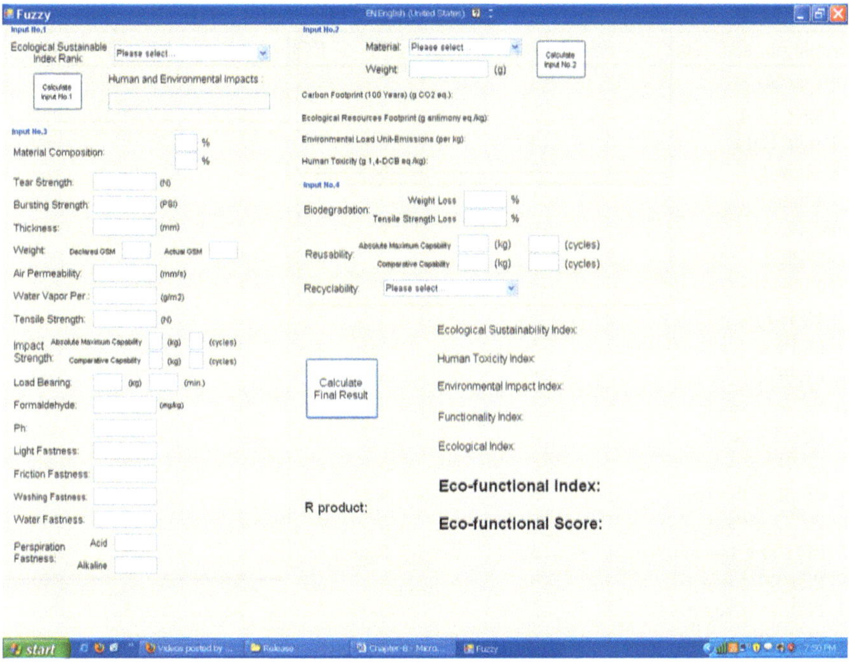

Fig. 7.6 Interface of eco-functional assessment [1]

functionality part. They earned better values in terms of raw materials, manufacturing processes (LCA indices chosen), and disposal values considered for this model. They earned better scores even for their ecological properties (due to better biodegradation and recyclability indices, though their reusability values are the lowest); hence they achieved the maximum index scores compared to all of their counterparts. The point to be noted is, If they were bestowed with higher functionality, which could also enable them to be reused many times, they would have met all the requirements of this eco-functional assessment and met the eco-functional requirements [1].

From the eco-functional assessments, it is clearly understood that eco-functional assessment needs to consider many aspects and a material should ideally satisfy all the requirements to the maximum, and if not, at least meet the average requirement. An eco-functional assessment should encompass all the parameters from the entire life cycle to ensure minimal environmental and human impacts.

Recently, norms are becoming very stringent day by day. They necessarily have to be pretty much strict to mitigate urgent environmental issues. Again, each and every aspect of functional, ecological and other aspects needs focused attention. None of the aspects can be overlooked, for instance, a quality parameter such as the discrepancy between the actual and declared GSM and fibre content calls for attention. This model includes serious consideration of almost all the essential eco-functional considerations. Since there are the currently available criteria to

Table 7.14 Eco-functional scores and assessment results [1]

Sample no. and name	ESI	HTI	EII	FI	ECI	EFI	EFS/25	$R_{Product}$
1. Paper 40 g	5	5	5	2	4	5	21/25	Fail
2. Paper 75 g	5	5	5	2	4	5	21/25	Fail
3. Paper 150 g	5	5	5	3	4	5	22/25	Fail
4. Cotton-1	4	3	3	5	4	4	19/25	Fail
5. Cotton-2	5	1	1	5	4	4	16/25	Fail
6. HDPE-1	2	5	5	2	4	4	18/25	Fail
7. HDPE-2	2	5	5	2	4	4	18/25	Fail
8. HDPE-3	2	5	5	2	4	4	18/25	Fail
9. LDPE-1	2	5	5	3	3	4	18/25	Fail
10. LDPE-2	2	5	5	3	4	4	19/25	Fail
11. LDPE-3	2	5	5	3	4	4	19/25	Fail
12. PP 40 g sewn	1	5	5	4	4	4	19/25	Fail
13. PP 75 g sewn	1	5	5	5	4	4	20/25	Fail
14. PP 100 g sewn	1	5	5	5	4	4	20/25	Fail
15. PP 40 g thermo	1	5	5	4	4	4	19/25	Fail
16. PP 75 g thermo	1	5	5	5	4	4	20/25	Fail
17. PP 100 g thermo	1	5	5	5	4	4	20/25	Fail
18. PET 40 g sewn	3	5	5	4	4	5	21/25	Fail
19. PET 75 g sewn	3	5	5	5	4	5	22/25	Fail
20. PET 100 g sewn	3	5	5	5	4	5	22/25	Fail
21. PET 40 g thermo	3	5	5	3	3	4	19/25	Fail
22. PET 75 g thermo	3	5	5	4	4	5	21/25	Fail
23. PET 100 g thermo	3	5	5	5	4	5	22/25	Fail

evaluate the status of shopping bags in terms of the aspects discussed in this study, the current research considered the 50 % value of the maximum scores obtained from the whole list of samples. In future, a huge database needs to be developed for the whole range of materials, including shopping bags, so that decisions can be made according to the requirements of society and the different views and arguments of various stakeholders. Probably the point of decision may be moved to 70 or 80 % in future, say in 10 years time from now to benefit the environment [1].

7.9 Concluding Remarks

This chapter discussed the important aspect of this book, i.e. eco-functional assessment of grocery shopping bags. The basic principle and detailed concept of eco-functional approach were discussed in this chapter along with the details of the theoretical framework developed for eco-functional assessment. This chapter also dealt with details of the developed model for eco-functional assessment and the simulation of eco-functional assessment. Detailed discussions on derivation of eco-functional index and the model was demonstrated to assess 23 types of grocery shopping bags in terms of eco-functional attributes and eco-functional index.

References

1. Muthu SS (2011) Eco-functional assessment of grocery shopping bags. PhD thesis, The Hong Kong Polytechnic University, Hong Kong
2. Muthu SS, Li Y, Hu J-Y, Mok P-Y, Lin M (2013) Modelling and quantification of eco-functional index: the concept and applications of eco-functional assessment. Ecol Ind 26:33–43

Chapter 8
Conclusions

Grocery shopping bags are one of the important entities in our daily lives. A wide array of raw materials and different manufacturing techniques are utilized to manufacture grocery shopping bags. Plastic (LDPE and HDPE), paper, cotton woven bags, polypropylene and polyester non-woven bags form the important grocery bag types. Grocery bags are classified into two major types namely single use and reusable. LDPE, HDPE and paper bags belong to single use bags. Cotton woven bags, polypropylene and polyester non-woven bags belong to reusable bags category.

Shopping bags demand significant attention in literature not only because they are indispensable part of our life, rather their environmental concerns are very huge. Though every product impacts our living planet in its entire life cycle from the cradle till grave stage, impacts from products like shopping bags are very high, since they represent the symbol of throw-away society. These products deserve significant attention and urgent action to reduce their eco-impacts. Measurement of eco-impacts is the first stage of reducing those impacts and the measurement of various parameters in the entire life cycle of shopping bags was discussed in detail in this entire book under different chapters.

Though traditional life cycle approach is well established and is one of the successful method widely use to assess the environmental impacts made by any product, influence of functionality on each phase of life cycle deserves crucial attention. This was deciphered with the aid of eco-functional approach defined by the authors, where the influence of functionality on consumer behaviour, manufacturing processes, which in turn influences the ecological impacts, is the key point of consideration in eco-functional assessment.

This book discussed the details of every single aspect of life cycle of shopping bags beginning from raw materials till the end-of-life phase. This book reviewed and discussed the contribution of authors in every phase of life cycle of shopping bags namely:

1. Quantification of environmental impact and ecological sustainability of different raw materials used for shopping bags;

S. S. Muthu and Y. Li, *Assessment of Environmental Impact by Grocery Shopping Bags*, EcoProduction, DOI: 10.1007/978-981-4560-20-7_8,

2. Detailed life cycle inventory of manufacturing processes of shopping bags and the discussion of hot-spots in the manufacturing processes;
3. Study of consumer behaviour of shopping bags in China, Hong Kong and India;
4. Quantification of life cycle impacts of shopping bags pertaining to Chinese consumers;
5. Influence of consumer behaviour of shopping bags in carbon and eco-footprints of shopping bags;
6. Biodegradation potential of grocery shopping bags by soil burial test;
7. Quantification of recyclability potential index of different raw materials used for shopping bags.

Concept of eco-functional approach along with the basic principles and the theoretical framework of eco-functional assessment were discussed in this book. An eco-functional model developed for various products including four inputs and five outputs was discussed in detail. With the aid of eco-functional model, eco-functional index of any product can be quantified apart from evaluating the eco-functional attributes of any product. 23 types of shopping bags were evaluated in terms of different attributes lead to eco-functional assessment and eventually those were assessed in the light of eco-functional attributes and their eco-functional scores/indices were derived. The developed model can be used to evaluate any product on earth; its applications were explored for grocery shopping bags. Assessment by eco-functional approach stresses the importance of every attribute of any product in terms of its environmental and human impacts.

Consumers must be aware of the eco-impacts of any product and shopping bags in particular and they must be told that if they reuse or recycle a bag to a certain level, how much impacts can be prevented. One of the main parameters of a shopping bag is maximum reusability, which is the critical parameter that decides the entire fate of a grocery bag in terms of eco-impacts. Now consumers do not have an idea that how many times a particular bag can be used in terms of its functional limits, keeping the fashion aspects aside. Scientific instruments like the eco-functional tester, developed by the authors must be utilized to quantify the eco-functional parameters of a grocery bag and they must be indicated in the bag along with the other details such as wash care instructions. Parameters involved in eco-functional assessment must be included in ascertaining the eco and human impacts of a bag and in simple terms such as the eco-functional index must be indicated in the bag. Educating the consumers in the light of all these issues will bring enormous benefits to our planet in terms of trimming down the potential impacts.

With regards to the reduction of eco-impacts of shopping bags, they need emergent action and it is the need of the hour to reduce the eco-impacts of shopping bags. Every individual and every citizen of any country knows about the eco-impacts made by shopping bags these days due to the awareness of various environmental issues happening around the globe in general and the severity of those issues is added by disposal of grocery shopping bags. Many governments have banned plastic bags usage and started levying impacts for the usage of plastic bags.

Due to this, reusable bags become popular and started replacing single use bags used for grocery purposes. It is the responsibility of a customer to use a grocery bag responsibly. As said earlier, if a reusable bag is not reused till its functional limit, its impact will be very high than the impacts created by single use bags. Consumer behaviour will play a major role in minimizing the eco-impacts of shopping bags.

Following the principles of 3R's (Reduce, Reuse and Recycle) in case of any product and shopping bags specifically will help to reduce the eco-impacts made by them. Reduction of waste and energy, resources in each life cycle stage to the maximum possible extent, reuse of the bags to the maximum possible extent and recycle them at the end-of-life will do wonders in terms of reduced eco-impacts. Reduction in every stage of life rather than focusing on one particular phase will yield better results in terms of reduced environmental impacts. Reduction of impacts pertaining to raw material extraction till the manufacturing phase lies in the hands of manufacturers, they can think of reducing it in many ways such as working with renewable energy sources. When it comes to use and disposal phases, it purely lies in the hands of customers, where responsibility of consumers in terms of using a bag till it looses its functional value and diverting it to reuse/recycling options instead of disposing at landfill brings the eco-impacts to a minimum level. Governmental support at this juncture in terms of encouraging recycling options and bringing recycling policies will do wonders in terms of reducing eco-impacts. It must be a joint venture by both individuals and governments to trim down the environmental impacts created by those shopping bags.

Annexure

```
// Test01.cpp: 主项目文件。

#include "stdafx.h"
#include "Form1.h"
#include <stdio.h>
using namespace System::Runtime::InteropServices;

using namespace Test01;

int Result_Combine;

int ESI_Index;

int HTI_Index;

int CFPI_Index;
int ERFPI_Index;
int ELUI_Index;
int EII;
int EII_Index;

int TenS_Index;
int TS_Index;
int BS_Index;
int SI;
int SI_Index;

int IR_Index;
```

S. S. Muthu and Y. Li, *Assessment of Environmental Impact by Grocery Shopping Bags*, EcoProduction, DOI: 10.1007/978-981-4560-20-7,

```
int Ph_Index;
int F_Index;
int HSI;
int HSI_Index;

int AP_Index;
int VP_Index;
int Pbl;
int Pbl_Index;

int LF_Index;
int FF_Index;
int WF_Index;
int WtF_Index;
int Ac_Index;
int Al_Index;
int CFI;
int CFI_Index;

int Q_Index;

int Ftn;
int Ftn_Index;

int B_Index;
int R_Index;
int Rec_Index;
int Eco;
int Eco_Index;

int EFS;


void Form1::Calculate_Click(System::Object^  sender,
System::EventArgs^  e)
{
                 int R_esir;

                 R_hei_TBox->Enabled = true;
                 R_hei_TBox->ReadOnly = true;

                 R_esir = R_esir_Combo->Items->IndexOf(R_esir_
```

```
Combo->Text);

                    ESI_Index = 0;

                    if(R_esir==1 || R_esir==2)
                        ESI_Index = 5;
                    else if(R_esir==3 || R_esir==4)
                        ESI_Index = 4;
                    else if(R_esir==5 || R_esir==6)
                        ESI_Index = 3;
                    else if(R_esir==7 || R_esir==8)
                        ESI_Index = 2;
                    else if(R_esir==9 || R_esir==10)
                        ESI_Index = 1;
                    else if(R_esir==11)
                        ESI_Index = 4;
                    else if(R_esir==12 || R_esir==13)
                        ESI_Index = 5;

                    if (R_esir==0)
                    {
                        MessageBox::Show("Please select
R_esir first.", "Error", MessageBoxButtons::OK,
MessageBoxIcon::Exclamation);
                        return;
                    }
                    else
                    {
                       switch  (R_esir)
                       {
                       case 1:
                           M_C_Text->Text = L"Organic Cotton";
                           M_C2_Text->Text = L"";
                           R_hei_TBox->Text = L"Close to None";
                           break;
                       case 2:
                           M_C_Text->Text = L"Flax or Paper";
                           M_C2_Text->Text = L"";
                           R_hei_TBox->Text = L"Very Less";
                           break;
                       case 3:
                           M_C_Text->Text = L"Cotton";
```

```
    M_C2_Text->Text = L"";
    R_hei_TBox->Text = L"Less";
    break;
case 4:
    M_C_Text->Text = L"Viscose";
    M_C2_Text->Text = L"";
    R_hei_TBox->Text = L"Moderately Less";
    break;
case 5:
    M_C_Text->Text = L"Wool";
    M_C2_Text->Text = L"";
    R_hei_TBox->Text = L"Moderate";
    break;
case 6:
    M_C_Text->Text = L"PET/Nylon 6";
    M_C2_Text->Text = L"";
    R_hei_TBox->Text = L"Moderately High";
    break;
case 7:
    M_C_Text->Text = L"Nylon 66";
    M_C2_Text->Text = L"";
    R_hei_TBox->Text = L"High";
    break;
case 8:
    M_C_Text->Text = L"LDPE/HDPE";
    M_C2_Text->Text = L"";
    R_hei_TBox->Text = L"Very High";
    break;
case 9:
    M_C_Text->Text = L"Polypropylene";
    M_C2_Text->Text = L"";
    R_hei_TBox->Text = L"Extreme";
    break;
case 10:
    M_C_Text->Text = L"Acrylic";
    M_C2_Text->Text = L"";
    R_hei_TBox->Text = L"Extremely High";
    break;
case 11:
    M_C_Text->Text = L"Cotton";
    M_C2_Text->Text = L"Polyester";
    R_hei_TBox->Text = L"Less";
```

```
                        break;
                    case 12:
                        M_C_Text->Text = L"Cotton";
                        M_C2_Text->Text = L"Polyester";
                        R_hei_TBox->Text = L"Less";
                        break;
                    case 13:
                        M_C_Text->Text = L"Cotton";
                        M_C2_Text->Text = L"Polyester";
                        R_hei_TBox->Text = L"Moderate";
                        break;
                }
            }
}

void Form1::R_esir_Combo_SelectedIndexChanged(System::Object^
sender, System::EventArgs^  e)
{
            R_hei_TBox->Enabled = false;
            R_hei_TBox->Text = L"";
}

void Form1::T_S_TBox_Click(System::Object^  sender,
System::EventArgs^  e)
{
    if(T_S_TBox->Text==L"")
        TS_Result->Text = L"(0.5~45)";
}

void Form1::T_S_TBox_TextChanged(System::Object^  sender,
System::EventArgs^  e)
{
    float value = -1;
    char* tmp = (char*)(void*)Marshal::StringToHGlobalAnsi(
T_S_TBox->Text );

    sscanf(tmp, "%f", &value);

    TS_Index = 0;
    if(value >=0.5 && value < 9.4)
        TS_Index = 1;
    else if(value >=9.4 && value < 18.3)
```

```
        TS_Index = 2;
    else if(value >=18.3 && value < 27.2)
        TS_Index = 3;
    else if(value >=27.2 && value < 36.1)
        TS_Index = 4;
    else if(value >=36.1 && value <= 45)
        TS_Index = 5;

    TS_Result->Visible = 0;
    if (TS_Index>=2.5)
        TS_Result->Text = L"Pass";
    else if (TS_Index<2.5 && TS_Index>0)
        TS_Result->Text = L"Fail";
    else
    {
        TS_Result->Text = L"(0.5~45)";
        TS_Result->Visible = 1;
    }
}

void Form1::B_S_TBox_Click(System::Object^  sender,
System::EventArgs^  e)
{
    if(B_S_TBox->Text==L"")
        BS_Result->Text = L"(11.9~126.7)";
}

void Form1::B_S_TBox_TextChanged(System::Object^  sender,
System::EventArgs^  e)
{
    float value = -1;
    char* tmp = (char*)(void*)Marshal::StringToHGlobalAnsi(
B_S_TBox->Text );

    sscanf(tmp, "%f", &value);

    BS_Index = 0;
    if(value >=11.9 && value < 34.86)
        BS_Index = 1;
    else if(value >=34.86 && value < 57.82)
        BS_Index = 2;
    else if(value >=57.82 && value < 80.78)
```

```
        BS_Index = 3;
    else if(value >=80.78 && value < 103.74)
        BS_Index = 4;
    else if(value >=103.74 && value <= 126.7)
        BS_Index = 5;

    BS_Result->Visible = 0;
    if (BS_Index>=2.5)
        BS_Result->Text = L"Pass";
    else if (BS_Index<2.5 && BS_Index>0)
        BS_Result->Text = L"Fail";
    else
    {
        BS_Result->Text = L"(11.9~126.7)";
        BS_Result->Visible = 1;
    }
}

void Form1::T_N_TBox_Click(System::Object^  sender,
System::EventArgs^  e)
{
    if(T_N_TBox->Text==L"")
        TN_Result->Text = L"(0.044~2.632)";
}

void Form1::T_N_TBox_TextChanged(System::Object^  sender,
System::EventArgs^  e)
{
    float value = -1;
    char* tmp = (char*)(void*)Marshal::StringToHGlobalAnsi(
T_N_TBox->Text );

    sscanf(tmp, "%f", &value);

    if(value >=0.044 && value <= 2.632)
        TN_Result->Visible = 0;
    else
    {
        TN_Result->Text = L"(0.044~2.632)";
        TN_Result->Visible = 1;
    }
}
```

```
void Form1::W_TBox_Click(System::Object^  sender,
System::EventArgs^  e)
{
}

void Form1::W_TBox_TextChanged(System::Object^  sender,
System::EventArgs^  e)
{
}

void Form1::WA_TBox_Click(System::Object^  sender,
System::EventArgs^  e)
{
}

void Form1::WA_TBox_TextChanged(System::Object^  sender,
System::EventArgs^  e)
{
    float value_W = -1;
    float value_WA = -1;
    float diff;

    char* tmp = (char*)(void*)Marshal::StringToHGlobalAnsi(
W_TBox->Text );

    sscanf(tmp, "%f", &value_W);

    tmp = (char*)(void*)Marshal::StringToHGlobalAnsi( WA_TBox-
>Text );

    sscanf(tmp, "%f", &value_WA);

    diff = value_WA - value_W;

    if(diff<0)
        diff = diff * -1;

    W_Result->Visible = 0;
    if(diff/value_W>0.1)
        W_Result->Text = L"Fail";
    else
```

```
        W_Result->Text = L"Pass";
}

void Form1::A_P_TBox_Click(System::Object^  sender,
System::EventArgs^  e)
{
    if(A_P_TBox->Text==L"")
        AP_Result->Text = L"(0~789)";
}

void Form1::A_P_TBox_TextChanged(System::Object^  sender,
System::EventArgs^  e)
{
    float value = -1;
    char* tmp = (char*)(void*)Marshal::StringToHGlobalAnsi(
A_P_TBox->Text );

    sscanf(tmp, "%f", &value);

    AP_Index = 0;
    if(value >=0 && value < 157.8)
        AP_Index = 1;
    else if(value >=157.8 && value < 315.6)
        AP_Index = 2;
    else if(value >=315.6 && value < 473.4)
        AP_Index = 3;
    else if(value >=473.4 && value < 631.2)
        AP_Index = 4;
    else if(value >=631.2 && value <= 789)
        AP_Index = 5;

    AP_Result->Visible = 0;
    if (AP_Index>=2.5)
        AP_Result->Text = L"Pass";
    else if (AP_Index<2.5 && AP_Index>0)
        AP_Result->Text = L"Fail";
    else
    {
        AP_Result->Text = L"(0~789)";
        AP_Result->Visible = 1;
    }
}
```

```
void Form1::V_P_TBox_Click(System::Object^  sender,
System::EventArgs^  e)
{
    if(V_P_TBox->Text==L"")
        VP_Result->Text = L"(14.3~1003.6)";
}

void Form1::V_P_TBox_TextChanged(System::Object^  sender,
System::EventArgs^  e)
{
    float value = -1;
    char* tmp = (char*)(void*)Marshal::StringToHGlobalAnsi(
V_P_TBox->Text );

    sscanf(tmp, "%f", &value);

    VP_Index = 0;
    if(value >=14.3 && value < 212.16)
        VP_Index = 1;
    else if(value >=212.16 && value < 410.02)
        VP_Index = 2;
    else if(value >=410.02 && value < 607.88)
        VP_Index = 3;
    else if(value >=607.88 && value < 805.74)
        VP_Index = 4;
    else if(value >=805.74 && value <= 1003.6)
        VP_Index = 5;

    VP_Result->Visible = 0;
    if (VP_Index>=2.5)
        VP_Result->Text = L"Pass";
    else if (VP_Index<2.5 && VP_Index>0)
        VP_Result->Text = L"Fail";
    else
    {
        VP_Result->Text = L"(14.3~1003.6)";
        VP_Result->Visible = 1;
    }
}

void Form1::W_P_TBox_Click(System::Object^  sender,
```

```
System::EventArgs^  e)
{
    if(W_P_TBox->Text==L"")
        WP_Result->Text = L"(60~766.1)";
}

void Form1::W_P_TBox_TextChanged(System::Object^  sender,
System::EventArgs^  e)
{
    float value = -1;
    char* tmp = (char*)(void*)Marshal::StringToHGlobalAnsi(
W_P_TBox->Text );

    sscanf(tmp, "%f", &value);

    TenS_Index = 0;
    if(value >=60 && value < 201.22)
        TenS_Index = 1;
    else if(value >=201.22 && value < 342.44)
        TenS_Index = 2;
    else if(value >=342.44 && value < 483.66)
        TenS_Index = 3;
    else if(value >=483.66 && value < 624.88)
        TenS_Index = 4;
    else if(value >=624.88 && value <= 766.1)
        TenS_Index = 5;

    WP_Result->Visible = 0;
    if (TenS_Index>=2.5)
        WP_Result->Text = L"Pass";
    else if (TenS_Index<2.5 && TenS_Index>0)
        WP_Result->Text = L"Fail";
    else
    {
        WP_Result->Text = L"(60~766.1)";
        WP_Result->Visible = 1;
    }
}

void Form1::Ph_TBox_Click(System::Object^  sender,
System::EventArgs^  e)
{
```

```
    if(Ph_TBox->Text==L"")
        Ph_Result->Text = L"(5.92~9.12)";
}

void Form1::Ph_TBox_TextChanged(System::Object^  sender,
System::EventArgs^  e)
{
    float value = -1;
    char* tmp = (char*)(void*)Marshal::StringToHGlobalAnsi(
Ph_TBox->Text );

    sscanf(tmp, "%f", &value);

    if(value != -1 && (value <4 || value >9))
        Ph_Index = 1;
    else if((value >= 4 && value <= 9))
        Ph_Index = 5;

    Ph_Result->Visible = 0;
    if (Ph_Index>=2.5)
        Ph_Result->Text = L"Pass";
    else if (Ph_Index<2.5 && Ph_Index>0)
        Ph_Result->Text = L"Fail";
    else
    {
        Ph_Result->Text = L"(5.92~9.12)";
        Ph_Result->Visible = 1;
    }
}

void Form1::MC_TBox_Click(System::Object^  sender,
System::EventArgs^  e)
{
    if(M_C_Text->Text==L"")
        MessageBox::Show("Please finish Input No.1 first.",
"Error", MessageBoxButtons::OK, MessageBoxIcon::Exclamation);
}

void Form1::MC2_TBox_Click(System::Object^  sender,
System::EventArgs^  e)
{
    if(M_C2_Text->Text==L"")
```

```
        MessageBox::Show("Please finish Input No.1 first.",
"Error", MessageBoxButtons::OK, MessageBoxIcon::Exclamation);
}

void Form1::MC_TBox_TextChanged(System::Object^  sender,
System::EventArgs^  e)
{
    int R_esir;
    float value = -1;

    Result_Combine=0;
    MC_Result->Visible = 0;

    char* tmp = (char*)(void*)Marshal::StringToHGlobalAnsi(
MC_TBox->Text );

    sscanf(tmp, "%f", &value);

    R_esir = R_esir_Combo->Items->IndexOf(R_esir_Combo->Text);
    if(R_esir>=1 && R_esir<=10)
    {
        if((value >= 95 && value <= 100))
            MC_Result->Text = L"Pass";
        else
            MC_Result->Text = L"Fail";
    }
    else if(R_esir==11)
    {
        if((value >= 76 && value <= 84))
            Result_Combine++;
    }
    else if(R_esir==12)
    {
        if((value >= 63.5 && value <= 70.5))
            Result_Combine++;
    }
    else if(R_esir==13)
    {
        if((value >= 47.5 && value <= 52.5))
            Result_Combine++;
    }
}
```

```
void Form1::MC2_TBox_TextChanged(System::Object^  sender,
System::EventArgs^  e)
{
    int R_esir;
    float value = -1;

    MC_Result->Visible = 0;

    char* tmp = (char*)(void*)Marshal::StringToHGlobalAnsi(
MC2_TBox->Text );

    sscanf(tmp, "%f", &value);

    R_esir = R_esir_Combo->Items->IndexOf(R_esir_Combo->Text);

    if(R_esir==11)
    {
        if((value >= 19 && value <= 21) && Result_Combine>0)
            MC_Result->Text = L"Pass";
        else
            MC_Result->Text = L"Fail";
    }
    else if(R_esir==12)
    {
        if((value >= 31 && value <= 35) && Result_Combine>0)
            MC_Result->Text = L"Pass";
        else
            MC_Result->Text = L"Fail";
    }
    else if(R_esir==13)
    {
        if((value >= 47.5 && value <= 52.5) && Result_
Combine>0)
            MC_Result->Text = L"Pass";
        else
            MC_Result->Text = L"Fail";
    }
}

void Form1::I_S_k1_TBox_Click(System::Object^  sender,
System::EventArgs^  e)
```

```
{
}

void Form1::I_S_k1_TBox_TextChanged(System::Object^  sender,
System::EventArgs^  e)
{
}

void Form1::I_S_c1_TBox_Click(System::Object^  sender,
System::EventArgs^  e)
{
}

void Form1::I_S_c1_TBox_TextChanged(System::Object^  sender,
System::EventArgs^  e)
{
}

void Form1::I_S_k2_TBox_Click(System::Object^  sender,
System::EventArgs^  e)
{
}

void Form1::I_S_k2_TBox_TextChanged(System::Object^  sender,
System::EventArgs^  e)
{
}

void Form1::I_S_c2_TBox_Click(System::Object^  sender,
System::EventArgs^  e)
{
}

//min 1, max 5
void Form1::I_S_c2_TBox_TextChanged(System::Object^  sender,
System::EventArgs^  e)
{
    float value = -1;
    char* tmp = (char*)(void*)Marshal::StringToHGlobalAnsi(
I_S_c2_TBox->Text );

    sscanf(tmp, "%f", &value);
```

```
    IR_Index = value;

    IS_Result->Visible = 0;
    if (IR_Index>=2.5)
        IS_Result->Text = L"Pass";
    else if (IR_Index<2.5 && IR_Index>0)
        IS_Result->Text = L"Fail";
}

void Form1::L_B_k_TBox_Click(System::Object^  sender,
System::EventArgs^  e)
{
}

//min 4, max 35
void Form1::L_B_k_TBox_TextChanged(System::Object^  sender,
System::EventArgs^  e)
{
    float value = -1;
    char* tmp = (char*)(void*)Marshal::StringToHGlobalAnsi(
L_B_k_TBox->Text );

    sscanf(tmp, "%f", &value);

    LB_Result->Visible = 0;
    if(value >=4 && value < 19.5)
        LB_Result->Text = L"Fail";
    else if((value >= 19.5 && value <= 35))
        LB_Result->Text = L"Pass";
}

void Form1::L_B_m_TBox_Click(System::Object^  sender,
System::EventArgs^  e)
{
}

void Form1::L_B_m_TBox_TextChanged(System::Object^  sender,
System::EventArgs^  e)
{
}
```

```
void Form1::F_TBox_Click(System::Object^  sender,
System::EventArgs^  e)
{
    if(F_TBox->Text==L"")
        F_Result->Text = L"(5.86~14.06)";
}

void Form1::F_TBox_TextChanged(System::Object^  sender,
System::EventArgs^  e)
{
    float value = -1;
    char* tmp = (char*)(void*)Marshal::StringToHGlobalAnsi(
F_TBox->Text );

    sscanf(tmp, "%f", &value);

    if(value >300)
        F_Index = 1;
    else if((value <= 300) && value !=-1)
        F_Index = 5;

    F_Result->Visible = 0;
    if (F_Index>=2.5)
        F_Result->Text = L"Pass";
    else if (F_Index<2.5 && F_Index>0)
        F_Result->Text = L"Fail";
    else
    {
        F_Result->Text = L"(5.86~14.06)";
        F_Result->Visible = 1;
    }
}

void Form1::L_F_TBox_Click(System::Object^  sender,
System::EventArgs^  e)
{
    if(L_F_TBox->Text==L"")
        LF_Result->Text = L"(1~5)";
}

void Form1::L_F_TBox_TextChanged(System::Object^  sender,
System::EventArgs^  e)
```

```
{
    float value = -1;
    char* tmp = (char*)(void*)Marshal::StringToHGlobalAnsi(
L_F_TBox->Text );

    sscanf(tmp, "%f", &value);

    //<3 Fail, 3 and >3 Pass(RHI) (RQ), 1~5
    LF_Result->Visible = 0;
    if(value >= 1 && value < 3)
        LF_Result->Text = L"Fail";
    else if(value>=3 && value <= 5)
        LF_Result->Text = L"Pass";
    else
    {
        LF_Result->Text = L"(1~5)";
        LF_Result->Visible = 1;
    }

    LF_Index = value;
}

void Form1::F_F_TBox_Click(System::Object^  sender,
System::EventArgs^  e)
{
    if(F_F_TBox->Text==L"")
        FF_Result->Text = L"(1~5)";
}

void Form1::F_F_TBox_TextChanged(System::Object^  sender,
System::EventArgs^  e)
{
    float value = -1;
    char* tmp = (char*)(void*)Marshal::StringToHGlobalAnsi(
F_F_TBox->Text );

    sscanf(tmp, "%f", &value);

    //<3 Fail, 3 and >3 Pass(RHI) (RQ), 1~5
    FF_Result->Visible = 0;
    if(value >= 1 && value < 3)
        FF_Result->Text = L"Fail";
```

```
    else if(value>=3 && value <= 5)
        FF_Result->Text = L"Pass";
    else
    {
        FF_Result->Text = L"(1~5)";
        FF_Result->Visible = 1;
    }

    FF_Index = value;
}

void Form1::W_F_TBox_Click(System::Object^  sender,
System::EventArgs^  e)
{
    if(W_F_TBox->Text==L"")
        WF_Result->Text = L"(1~5)";
}

void Form1::W_F_TBox_TextChanged(System::Object^  sender,
System::EventArgs^  e)
{
    float value = -1;
    char* tmp = (char*)(void*)Marshal::StringToHGlobalAnsi(
W_F_TBox->Text );

    sscanf(tmp, "%f", &value);

    //<3 Fail, 3 and >3 Pass(RHI) (RQ), 1~5
    WF_Result->Visible = 0;
    if(value >=1 && value < 3)
        WF_Result->Text = L"Fail";
    else if(value>=3 && value <= 5)
        WF_Result->Text = L"Pass";
    else
    {
        WF_Result->Text = L"(1~5)";
        WF_Result->Visible = 1;
    }

    WF_Index = value;
}
```

```
void Form1::Wt_F_TBox_Click(System::Object^  sender,
System::EventArgs^  e)
{
    if(Wt_F_TBox->Text==L"")
        WtF_Result->Text = L"(1~5)";
}

void Form1::Wt_F_TBox_TextChanged(System::Object^  sender,
System::EventArgs^  e)
{
    float value = -1;
    char* tmp = (char*)(void*)Marshal::StringToHGlobalAnsi(
Wt_F_TBox->Text );

    sscanf(tmp, "%f", &value);

    //<3 Fail, 3 and >3 Pass(RHI) (RQ), 1~5
    WtF_Result->Visible = 0;
    if(value >= 1 && value < 3)
        WtF_Result->Text = L"Fail";
    else if(value>=3 && value <= 5)
        WtF_Result->Text = L"Pass";
    else
    {
        WtF_Result->Text = L"(1~5)";
        WtF_Result->Visible = 1;
    }

    WtF_Index = value;
}

void Form1::Ac_TBox_TextChanged(System::Object^  sender,
System::EventArgs^  e)
{
    float value_Ac = -1;
    float value_Al = -1;
    char* tmp;
    tmp = (char*)(void*)Marshal::StringToHGlobalAnsi( Ac_TBox-
>Text );
    sscanf(tmp, "%f", &value_Ac);
    tmp = (char*)(void*)Marshal::StringToHGlobalAnsi( Al_TBox-
>Text );
```

```
    sscanf(tmp, "%f", &value_Al);

    //<3 Fail, 3 and >3 Pass(RHI) (RQ), 1~5
    PF_Result->Visible = 0;
    if((value_Ac >= 1 && value_Ac < 3) || (value_Al >= 1 &&
value_Al < 3))
        PF_Result->Text = L"Fail";
    else if((value_Ac >=3 && value_Ac <= 5) && (value_Al >=3 &&
value_Al <= 5))
        PF_Result->Text = L"Pass";
    else
    {
        PF_Result->Text = L"(1~5)";
        PF_Result->Visible = 1;
    }

    Ac_Index = value_Ac;
}

void Form1::Al_TBox_TextChanged(System::Object^  sender,
System::EventArgs^  e)
{
    float value_Ac = -1;
    float value_Al = -1;
    char* tmp;
    tmp = (char*)(void*)Marshal::StringToHGlobalAnsi( Ac_TBox-
>Text );
    sscanf(tmp, "%f", &value_Ac);
    tmp = (char*)(void*)Marshal::StringToHGlobalAnsi( Al_TBox-
>Text );
    sscanf(tmp, "%f", &value_Al);

    //<3 Fail, 3 and >3 Pass(RHI) (RQ), 1~5
    PF_Result->Visible = 0;
    if((value_Ac >= 1 && value_Ac < 3) || (value_Al >= 1 &&
value_Al < 3))
        PF_Result->Text = L"Fail";
    else if((value_Ac >=3 && value_Ac <= 5) && (value_Al >=3 &&
value_Al <= 5))
        PF_Result->Text = L"Pass";
    else
    {
```

```
        PF_Result->Text = L"(1~5)";
        PF_Result->Visible = 1;
    }

    Al_Index = value_Al;
}

void Form1::W_L_TBox_TextChanged(System::Object^  sender,
System::EventArgs^  e)
{
    float value = -1;
    char* tmp;
    tmp = (char*)(void*)Marshal::StringToHGlobalAnsi( W_L_TBox-
>Text );
    sscanf(tmp, "%f", &value);

    B_Index = 0;
    if(value >=0 && value <=20)
        B_Index = 1;
    else if(value >20 && value <=40)
        B_Index = 2;
    else if(value >40 && value <=60)
        B_Index = 3;
    else if(value >60 && value <=80)
        B_Index = 4;
    else if(value >80 && value <= 100)
        B_Index = 5;

    B_Result->Visible = 0;
    if (B_Index>=2.5)
        B_Result->Text = L"Pass";
    else if (B_Index<2.5 && B_Index>0)
        B_Result->Text = L"Fail";
}

void Form1::T_L_TBox_TextChanged(System::Object^  sender,
System::EventArgs^  e)
{
}

void Form1::R_k1_TBox_Click(System::Object^  sender,
System::EventArgs^  e)
```

```
{
}

void Form1::R_k1_TBox_TextChanged(System::Object^  sender,
System::EventArgs^  e)
{
}

void Form1::R_c1_TBox_Click(System::Object^  sender,
System::EventArgs^  e)
{
}

void Form1::R_c1_TBox_TextChanged(System::Object^  sender,
System::EventArgs^  e)
{
}

void Form1::R_k2_TBox_Click(System::Object^  sender,
System::EventArgs^  e)
{
}

void Form1::R_k2_TBox_TextChanged(System::Object^  sender,
System::EventArgs^  e)
{
}

void Form1::R_c2_TBox_Click(System::Object^  sender,
System::EventArgs^  e)
{
}

//min 0, max 500
void Form1::R_c2_TBox_TextChanged(System::Object^  sender,
System::EventArgs^  e)
{
    int value = -1;
    char* tmp = (char*)(void*)Marshal::StringToHGlobalAnsi(
R_c2_TBox->Text );

    sscanf(tmp, "%d", &value);
```

```
    R_Index = 0;
    if(value >=0 && value <= 100)
        R_Index = 1;
    else if(value >100 && value <= 200)
        R_Index = 2;
    else if(value >200 && value <= 300)
        R_Index = 3;
    else if(value >300 && value <= 400)
        R_Index = 4;
    else if(value >400)
        R_Index = 5;

    R_Result->Visible = 0;
    if (R_Index>=2.5)
        R_Result->Text = L"Pass";
    else if (R_Index<2.5 && R_Index>0)
        R_Result->Text = L"Fail";
}

void Form1::Rec_CBox_SelectedIndexChanged(System::Object
^sender, System::EventArgs ^e)
{
    int index;

    index = Rec_CBox->Items->IndexOf(Rec_CBox->Text);

    switch (index)
    {
    case 0:
        Rec_Result->Text = L"";
        break;
    case 1: //PP
        Rec_Index = 5;
        break;
    case 2: //Polyester
        Rec_Index = 5;
        break;
    case 3: //HDPE
        Rec_Index = 5;
        break;
    case 4: //LDPE
```

```
            Rec_Index = 4;
            break;
        case 5: //Paper
            Rec_Index = 4;
            break;
        case 6: //Acrylic
            Rec_Index = 3;
            break;
        case 7: //80% Cotton, 20% PET
            Rec_Index = 3;
            break;
        case 8: //67% Cotton, 33% PET
            Rec_Index = 3;
            break;
        case 9: //50% Cotton, 50% PET
            Rec_Index = 3;
            break;
        case 10:    //Cotton
            Rec_Index = 3;
            break;
        case 11:    //Wool
            Rec_Index = 2;
            break;
        case 12:    //Viscose
            Rec_Index = 2;
            break;
        case 13:    //Nylon 6
            Rec_Index = 1;
            break;
        case 14:    //Nylon 66
            Rec_Index = 1;
            break;
    }

    Rec_Result->Visible = 0;
    if (Rec_Index>=2.5)
        Rec_Result->Text = L" Pass" ;
    else if (Rec_Index<2.5 && Rec_Index>0)
        Rec_Result->Text = L" Fail" ;
}

void Form1::Mat_TBox_Click(System::Object^  sender,
```

```
System::EventArgs^  e)
{
    if(Mat_CBox->Items->IndexOf(Mat_CBox->Text)==0)
        MessageBox::Show( "Please select material first." ,
 "Error" , MessageBoxButtons::OK, MessageBoxIcon::Exclamation);
    else
    {
        Mat_TBox->Text = L" " ;
        CF_R_Text->Text = L" " ;
        ER_R_Text->Text = L" " ;
        EL_R_Text->Text = L" " ;
        HT_R_Text->Text = L" " ;
        CF_Result->Text = L" " ;
        ER_Result->Text = L" " ;
        EL_Result->Text = L" " ;
        HT_Result->Text = L" " ;
    }
}

void Form1::Mat_CBox_SelectedIndexChanged(System::Object
^sender, System::EventArgs ^e)
{
    Mat_TBox->Text = L" " ;
}

void Form1::Calculate2_Click(System::Object ^sender,
System::EventArgs ^e)
{
    if(Mat_CBox->Items->IndexOf(Mat_CBox->Text)==0)
    {
        MessageBox::Show( "Please select material first." ,
 "Error" , MessageBoxButtons::OK, MessageBoxIcon::Exclamation);
        return;
    }

    int index;
    float value = 0;
    float CF[7] = {0, 19.8285, 9.46899061, 53.3217988,
40.1943222, 24.6409392, 26.2830375};
    float ER[7] = {0, 0.000195, 0.0000977, 0.000494, 0.000415,
0.000243, 0.000256};
    float EL[7] = {0, 0.00128184, 0.0006126, 0.00344478,
```

```
0.00260043, 0.00159281, 0.00170225};
    float HT[7] = {0, 11.4334133, 5.74131101, 29.8053522,
24.5462048, 14.2831446, 15.1055028};
    float CF_R, ER_R, EL_R, HT_R;

    char* tmp = (char*)(void*)Marshal::StringToHGlobalAnsi(
Mat_TBox->Text );
    sscanf(tmp, "%f", &value);

    index = Mat_CBox->Items->IndexOf(Mat_CBox->Text);

    CF_R = value * CF[index];
    ER_R = value * ER[index];
    EL_R = value * EL[index];
    HT_R = value * HT[index];

    CFPI_Index = 0;
    if(CF_R>=118 && CF_R<2024.4)
        CFPI_Index = 5;
    else if(CF_R>=2024.4 && CF_R<3930.8)
        CFPI_Index = 4;
    else if(CF_R>=3930.8 && CF_R<5837.2)
        CFPI_Index = 3;
    else if(CF_R>=5837.2 && CF_R<7743.6)
        CFPI_Index = 2;
    else if(CF_R>=7743.6 && CF_R<=9650)
        CFPI_Index = 1;

    CF_R_Text->Visible = 0;
    if (CFPI_Index>=2.5)
        CF_R_Text->Text = L"Pass";
    else if (CFPI_Index<2.5 && CFPI_Index>0)
        CF_R_Text->Text = L"Fail";
    else
    {
        CF_R_Text->Text = L"(118~9650)";
        CF_R_Text->Visible = 1;
    }


    ERFPI_Index = 0;
    if(ER_R>=0.00117 && ER_R<0.020936)
```

```
    ERFPI_Index = 5;
else if(ER_R>=0.020936 && ER_R<0.040702)
    ERFPI_Index = 4;
else if(ER_R>=0.040702 && ER_R<0.060468)
    ERFPI_Index = 3;
else if(ER_R>=0.060468 && ER_R<0.080234)
    ERFPI_Index = 2;
else if(ER_R>=0.080234 && ER_R<=0.1)
    ERFPI_Index = 1;

ER_R_Text->Visible = 0;
if (ERFPI_Index>=2.5)
    ER_R_Text->Text = L"Pass";
else if (ERFPI_Index<2.5 && ERFPI_Index>0)
    ER_R_Text->Text = L"Fail";
else
{
    ER_R_Text->Text = L"(0.00117~0.1)";
    ER_R_Text->Visible = 1;
}


ELUI_Index = 0;
if(EL_R>=0.0076 && EL_R<0.13208)
    ELUI_Index = 5;
else if(EL_R>=0.13208 && EL_R<0.25656)
    ELUI_Index = 4;
else if(EL_R>=0.25656 && EL_R<0.38104)
    ELUI_Index = 3;
else if(EL_R>=0.38104 && EL_R<0.50552)
    ELUI_Index = 2;
else if(EL_R>=0.50552 && EL_R<=0.63)
    ELUI_Index = 1;

EL_R_Text->Visible = 0;
if (ELUI_Index>=2.5)
    EL_R_Text->Text = L"Pass";
else if (ELUI_Index<2.5 && ELUI_Index>0)
    EL_R_Text->Text = L"Fail";
else
{
    EL_R_Text->Text = L"(0.0076~0.63)";
```

```
    EL_R_Text->Visible = 1;
}


EII = CFPI_Index + ERFPI_Index + ELUI_Index;

if(EII>=1 && EII<=3)
    EII_Index = 1;
else if(EII>=4 && EII<=6)
    EII_Index = 2;
else if(EII>=7 && EII<=9)
    EII_Index = 3;
else if(EII>=10 && EII<=12)
    EII_Index = 4;
else if(EII>=13 && EII<=15)
    EII_Index = 5;


HTI_Index = 0;
if(HT_R>=68.6 && HT_R<1233.1)
    HTI_Index = 5;
else if(HT_R>=1233.1 && HT_R<2397.6)
    HTI_Index = 4;
else if(HT_R>=2397.6 && HT_R<3562.1)
    HTI_Index = 3;
else if(HT_R>=3562.1 && HT_R<4726.6)
    HTI_Index = 2;
else if(HT_R>=4726.6 && HT_R<=5891.1)
    HTI_Index = 1;

HT_R_Text->Visible = 0;
if (HTI_Index>=2.5)
    HT_R_Text->Text = L"Pass";
else if (HTI_Index<2.5 && HTI_Index>0)
    HT_R_Text->Text = L"Fail";
else
{
    HT_R_Text->Text = L"(68.6~5891.1)";
    HT_R_Text->Visible = 1;
}
```

```
    sprintf(tmp, “%f”, CF_R);
    CF_Result->Text = Marshal::PtrToStringAnsi((IntPtr) (char
*)tmp);

    sprintf(tmp, “%f”, ER_R);
    ER_Result->Text = Marshal::PtrToStringAnsi((IntPtr) (char
*)tmp);

    sprintf(tmp, “%f”, EL_R);
    EL_Result->Text = Marshal::PtrToStringAnsi((IntPtr) (char
*)tmp);

    sprintf(tmp, “%f”, HT_R);
    HT_Result->Text = Marshal::PtrToStringAnsi((IntPtr) (char
*)tmp);

}

void Form1::Calculate_Final_Click(System::Object ^sender,
System::EventArgs ^e)
{
    if(
        R_hei_TBox->Text == L” ”  ||
        T_S_TBox->Text == L” ”  ||
        TS_Result->Text == L” ”  ||
        B_S_TBox->Text == L” ”  ||
        BS_Result->Text == L” ”  ||
        T_N_TBox->Text == L” ”  ||
        TN_Result->Text == L” ”  ||
        W_TBox->Text == L” ”  ||
        W_Result->Text == L” ”  ||
        A_P_TBox->Text == L” ”  ||
        AP_Result->Text == L” ”  ||
        V_P_TBox->Text == L” ”  ||
        VP_Result->Text == L” ”  ||
        W_P_TBox->Text == L” ”  ||
        WP_Result->Text == L” ”  ||
        Ph_TBox->Text == L” ”  ||
        Ph_Result->Text == L” ”  ||
        MC_Result->Text == L” ”  ||
        I_S_k1_TBox->Text == L” ”  ||
        IS_Result->Text == L” ”  ||
```

```
        L_B_k_TBox->Text == L" "  ||
        L_B_m_TBox->Text == L" "  ||
        LB_Result->Text == L" "  ||
        F_TBox->Text == L" "  ||
        F_Result->Text == L" "  ||
        L_F_TBox->Text == L" "  ||
        LF_Result->Text == L" "  ||
        F_F_TBox->Text == L" "  ||
        FF_Result->Text == L" "  ||
        W_F_TBox->Text == L" "  ||
        WF_Result->Text == L" "  ||
        Wt_F_TBox->Text == L" "  ||
        WtF_Result->Text == L" "  ||
        Ac_TBox->Text == L" "  ||
        Al_TBox->Text == L" "  ||
        W_L_TBox->Text == L" "  ||
        T_L_TBox->Text == L" "  ||
        R_c2_TBox->Text == L" "  ||
        I_S_c2_TBox->Text == L" "  ||
        L_B_k_TBox->Text == L" "  ||
        R_Result->Text == L" "  ||
        Rec_Result->Text == L" "
        )
    {
        MessageBox::Show( "Please compleate all four
inputs first." ,  "Error" , MessageBoxButtons::OK,
MessageBoxIcon::Exclamation);
        return;
    }

    int Rq, Rf, R3rs, Rei, Rhi;      //0 fail; 1 pass;
    int Rqf, Reif;                            //0 poor; 1
average; 2 good
    int Rp;                                        //0 fail; 1
medium; 2 pass
    int count_Rq, count_Rf;
    count_Rq = 0;
    count_Rf = 0;

    if (
        WtF_Result->Text == L" Pass" //Colour fastness to
water pass
```

```
        )
        count_Rq++;
    if (
        PF_Result->Text == L" Pass"       //Colour fastness to
perspiration ( Acid and Alkali) pass
        )
        count_Rq++;
    if (
        WF_Result->Text == L" Pass"       //Colour fastness to
washing pass
        )
        count_Rq++;
    if (
        FF_Result->Text == L" Pass"       //Colour fastness to
friction pass
        )
        count_Rq++;
    if (
        MC_Result->Text == L" Pass"       //Material
Composition / Fibre Content pass
        )
        count_Rq++;
    if (
        TN_Result->Text == L" Pass"       //Thickness pass
        )
        count_Rq++;
    if (
        W_Result->Text == L" Pass"        //Weight pass
        )
        count_Rq++;

    if (count_Rq>=4)
            Rq = 1;
    else
            Rq = 0;

    if (
        WP_Result->Text == L" Pass"       //Tensile strength
pass
        )
        count_Rf++;
    if (
```

```
        TS_Result->Text == L" Pass"        //Tear strength pass
        )
        count_Rf++;
    if (
        BS_Result->Text == L" Pass"        //Bursting strength
pass
        )
        count_Rf++;
    if (
        IS_Result->Text == L" Pass"        //Impact Strength
pass
        )
        count_Rf++;
    if (
        LB_Result->Text == L" Pass"        //Load Bearing pass
        )
        count_Rf++;
    if (
        VP_Result->Text == L" Pass"        //Water Vapour
Permeability pass
        )
        count_Rf++;

    if (count_Rf>=3)
            Rf = 1;
    else
            Rf = 0;

    if (
        R_Result->Text == L" Pass"              //Reusability
pass
        && Rec_Result->Text == L" Pass"        //Recyclability
pass
        )
            R3rs = 1;
    else
            R3rs = 0;

    if (
            (  R_hei_TBox->Text == L" Close to None"
            || R_hei_TBox->Text == L" Very Less"
            || R_hei_TBox->Text == L" Less"
```

```
            || R_hei_TBox->Text == L” Moderately Less”
            || R_hei_TBox->Text == L” Moderate” )
//REI is Close to none, Very less, less, moderately less and
moderate
        && B_Result->Text == L” Pass”
    //Biodegradation of material is pass
        && CF_R_Text->Text == L” Pass”
//Carbon footprint pass
        && ER_R_Text->Text == L” Pass”
//Ecological Resources footprint pass
        && EL_R_Text->Text == L” Pass”
//Environmental Load Unit-Emissions Pass
        )
            Rei = 1;
    else
            Rei = 0;

    if (
            (  R_hei_TBox->Text == L” Close to None”
            || R_hei_TBox->Text == L” Very Less”
            || R_hei_TBox->Text == L” Less”
            || R_hei_TBox->Text == L” Moderately Less”
            || R_hei_TBox->Text == L” Moderate” )
    //RHI is Close to none, Very less, less, moderately less
and moderate
        && Ph_Result->Text == L” Pass”
    //Ph Pass
        && F_Result->Text == L” Pass”
        //Formaldehyde Pass
        && B_Result->Text == L” Pass”
        //Biodegradation of material is pass
        && HT_R_Text->Text == L” Pass”
    //Human Toxicity Pass
        )
            Rhi = 1;
    else
            Rhi = 0;

    //Rqf from Rq, Rf
    if (Rq==1 && Rf==1)
        Rqf = 2;
    else if (Rq==1 && Rf==0)
```

```
    Rqf = 0;
else if (Rq==0 && Rf==1)
    Rqf = 1;
else if (Rq==0 && Rf==0)
    Rqf = 0;

//Reif from R3rs, Rei, Rhi
if (R3rs==1 && Rei==1 && Rhi==1)
    Reif = 2;
else if (R3rs==1 && Rei==1 && Rhi==0)
    Reif = 1;
else if (R3rs==1 && Rei==0 && Rhi==1)
    Reif = 1;
else if (R3rs==0 && Rei==1 && Rhi==1)
    Reif = 1;
else if (R3rs==1 && Rei==0 && Rhi==0)
    Reif = 0;
else if (R3rs==0 && Rei==1 && Rhi==0)
    Reif = 0;
else if (R3rs==0 && Rei==0 && Rhi==1)
    Reif = 0;
else if (R3rs==0 && Rei==0 && Rhi==0)
    Reif = 0;

//Rp from Rqf, Reif
if (Rqf==2 && Reif==2)
    Rp = 2;
else if (Rqf==2 && Reif==1)
    Rp = 2;
else if (Rqf==1 && Reif==2)
    Rp = 2;
else if (Rqf==1 && Reif==1)
    Rp = 1;
else if (Rqf==0 && Reif==0)
    Rp = 0;
else if (Rqf==0 && Reif==1)
    Rp = 0;
else if (Rqf==0 && Reif==2)
    Rp = 0;
else if (Rqf==2 && Reif==0)
    Rp = 0;
else if (Rqf==1 && Reif==0)
```

```
    Rp = 0;

if (Rp==2)
    Final_Result->Text = L" Pass" ;
else if (Rp==1)
    Final_Result->Text = L" Medium" ;
else if (Rp==0)
    Final_Result->Text = L" Fail" ;

if(ESI_Index==5)
    ESI_Result->Text = L" 5" ;
else if(ESI_Index==4)
    ESI_Result->Text = L" 4" ;
else if(ESI_Index==3)
    ESI_Result->Text = L" 3" ;
else if(ESI_Index==2)
    ESI_Result->Text = L" 2" ;
else if(ESI_Index==1)
    ESI_Result->Text = L" 1" ;
else
    ESI_Result->Text = L" " ;

if(HTI_Index==5)
    HTI_Result->Text = L" 5" ;
else if(HTI_Index==4)
    HTI_Result->Text = L" 4" ;
else if(HTI_Index==3)
    HTI_Result->Text = L" 3" ;
else if(HTI_Index==2)
    HTI_Result->Text = L" 2" ;
else if(HTI_Index==1)
    HTI_Result->Text = L" 1" ;
else
    HTI_Result->Text = L" " ;

if(EII_Index==5)
    EII_Result->Text = L" 5" ;
else if(EII_Index==4)
    EII_Result->Text = L" 4" ;
else if(EII_Index==3)
    EII_Result->Text = L" 3" ;
else if(EII_Index==2)
```

```
        EII_Result->Text = L” 2” ;
    else if(EII_Index==1)
        EII_Result->Text = L” 1” ;
    else
        EII_Result->Text = L” ” ;

    SI = TenS_Index + TS_Index + BS_Index;
    if(SI>=1 && SI<=3)
        SI_Index = 1;
    else if(SI>=4 && SI<=6)
        SI_Index = 2;
    else if(SI>=7 && SI<=9)
        SI_Index = 3;
    else if(SI>=10 && SI<=12)
        SI_Index = 4;
    else if(SI>=13 && SI<=15)
        SI_Index = 5;

    HSI = Ph_Index + F_Index;
    HSI_Index = HSI/2;

    Pbl = AP_Index + VP_Index;
    if(Pbl>=1 && Pbl<=2)
        Pbl_Index = 1;
    else if(Pbl>=3 && Pbl<=4)
        Pbl_Index = 2;
    else if(Pbl>=5 && Pbl<=6)
        Pbl_Index = 3;
    else if(Pbl>=7 && Pbl<=8)
        Pbl_Index = 4;
    else if(Pbl>=9 && Pbl<=10)
        Pbl_Index = 5;

    CFI = LF_Index + FF_Index + WF_Index + WtF_Index + Ac_Index
+ Al_Index;
    if(CFI<=10)
        CFI_Index = 1;
    else if(CFI>=11 && CFI<=15)
        CFI_Index = 2;
    else if(CFI>=16 && CFI<=20)
        CFI_Index = 3;
    else if(CFI>=21 && CFI<=25)
```

```
        CFI_Index = 4;
    else if(CFI>=26 && CFI<=30)
        CFI_Index = 5;

    Q_Index = 0;
    if(MC_Result->Text == L" Pass"  && W_Result->Text ==
L" Pass" )
        Q_Index = 5;
    else if(MC_Result->Text == L" Fail"  || W_Result->Text ==
L" Fail" )
        Q_Index = 1;

    Ftn = SI_Index + IR_Index + HSI_Index + Pbl_Index + CFI_
Index + Q_Index;
    if(Ftn>=6 && Ftn<=10)
        Ftn_Index = 1;
    else if(Ftn>=11 && Ftn<=15)
        Ftn_Index = 2;
    else if(Ftn>=16 && Ftn<=20)
        Ftn_Index = 3;
    else if(Ftn>=21 && Ftn<=25)
        Ftn_Index = 4;
    else if(Ftn>=26 && Ftn<=30)
        Ftn_Index = 5;

    if(Ftn_Index==5)
        FI_Result->Text = L" 5" ;
    else if(Ftn_Index==4)
        FI_Result->Text = L" 4" ;
    else if(Ftn_Index==3)
        FI_Result->Text = L" 3" ;
    else if(Ftn_Index==2)
        FI_Result->Text = L" 2" ;
    else if(Ftn_Index==1)
        FI_Result->Text = L" 1" ;
    else
        FI_Result->Text = L" " ;

    Eco = B_Index + R_Index + Rec_Index;
    if(Eco>0 && Eco<=3)
        Eco_Index = 1;
    else if(Eco>=4 && Eco<=6)
```

```
        Eco_Index = 2;
    else if(Eco>=7 && Eco<=9)
        Eco_Index = 3;
    else if(Eco>=10 && Eco<=12)
        Eco_Index = 4;
    else if(Eco>=13 && Eco<=15)
        Eco_Index = 5;

    if(Eco_Index==5)
        EI_Result->Text = L" 5" ;
    else if(Eco_Index==4)
        EI_Result->Text = L" 4" ;
    else if(Eco_Index==3)
        EI_Result->Text = L" 3" ;
    else if(Eco_Index==2)
        EI_Result->Text = L" 2" ;
    else if(Eco_Index==1)
        EI_Result->Text = L" 1" ;
    else
        EI_Result->Text = L" " ;

    EFS = ESI_Index + HTI_Index + EII_Index + Ftn_Index + Eco_
Index;
    if(EFS>=21 && EFS<=25)
        EFI_Result->Text = L" 5" ;
    else if(EFS>=16 && EFS<=20)
        EFI_Result->Text = L" 4" ;
    else if(EFS>=11 && EFS<=15)
        EFI_Result->Text = L" 3" ;
    else if(EFS>=6 && EFS<=10)
        EFI_Result->Text = L" 2" ;
    else if(EFS>0 && EFS<=5)
        EFI_Result->Text = L" 1" ;
    else
        EFI_Result->Text = L" " ;

    switch (EFS)
    {
    case 0:
        EFS_Result->Text = L" 0/25" ;
        break;
    case 1:
```

```
        EFS_Result->Text = L" 1/25" ;
        break;
    case 2:
        EFS_Result->Text = L" 2/25" ;
        break;
    case 3:
        EFS_Result->Text = L" 3/25" ;
        break;
    case 4:
        EFS_Result->Text = L" 4/25" ;
        break;
    case 5:
        EFS_Result->Text = L" 5/25" ;
        break;
    case 6:
        EFS_Result->Text = L" 6/25" ;
        break;
    case 7:
        EFS_Result->Text = L" 7/25" ;
        break;
    case 8:
        EFS_Result->Text = L" 8/25" ;
        break;
    case 9:
        EFS_Result->Text = L" 9/25" ;
        break;
    case 10:
        EFS_Result->Text = L" 10/25" ;
        break;
    case 11:
        EFS_Result->Text = L" 11/25" ;
        break;
    case 12:
        EFS_Result->Text = L" 12/25" ;
        break;
    case 13:
        EFS_Result->Text = L" 13/25" ;
        break;
    case 14:
        EFS_Result->Text = L" 14/25" ;
        break;
    case 15:
```

```
        EFS_Result->Text = L” 15/25” ;
        break;
    case 16:
        EFS_Result->Text = L” 16/25” ;
        break;
    case 17:
        EFS_Result->Text = L” 17/25” ;
        break;
    case 18:
        EFS_Result->Text = L” 18/25” ;
        break;
    case 19:
        EFS_Result->Text = L” 19/25” ;
        break;
    case 20:
        EFS_Result->Text = L” 20/25” ;
        break;
    case 21:
        EFS_Result->Text = L” 21/25” ;
        break;
    case 22:
        EFS_Result->Text = L” 22/25” ;
        break;
    case 23:
        EFS_Result->Text = L” 23/25” ;
        break;
    case 24:
        EFS_Result->Text = L” 24/25” ;
        break;
    case 25:
        EFS_Result->Text = L” 25/25” ;
        break;
    }

}


[STAThreadAttribute]
int main(array<System::String ^> ^args)
{
    // 在创建任何控件之前启用 Windows XP 可视化效果
```

```
    Application::EnableVisualStyles();
    Application::SetCompatibleTextRenderingDefault(false);

    // 创建主窗口并运行它
    Application::Run(gcnew Form1());

    return 0;
}
```

Author Biographies

Dr Subramanian Senthilkannan Muthu is working for SGS Hong Kong for its Global Sustainability Services as Eco-design Consultant after his doctoral research study from The Hong Kong Polytechnic University, Hong Kong. He finished his PhD dissertation entitled, "Eco-functional assessment of grocery shopping bags" with 14 scientific peer-reviewed scientific journal publications, 2 patents and 11 conference publications based on his dissertation to his credit. In total, he has authored more than 75 journal publications and numerous conference publications, 4 books and 2 book of chapters. He serves various peer-reviewed journals in environmental and textile sciences as an editor/editorial board member and also as a reviewer.

Dr. Yi Li is a full professor and director of Textile Bioengineering Research Center in the Institute of Textiles and Clothing, the Hong Kong Polytechnic University. He is a Fellow of the Textile Institute, Life Fellows of International Biographical Association and the Royal Society of Art, Commerce and Manufacturing, and adjunct professors of a number of universities in China and a member of several professional bodies. He is the Chairman of Textile Bioengineering and Informatics Society and the Editor-in-Chief of "Journal of Fiber Bioengineering and Informatics". By securing more than HK$90 million research funding and obtaining support from HK government, Guangdong government and industry, he established the Textile Bioengineering Research Center (TBRC) as an international platform of textile bioengineering to promote industry-university cooperation and accelerate technology transfer. TBRC is composed of 6 research laboratories: Textile Biomaterial Research Laboratory, Textile Biomaterial Testing Research Lab, Digital Apparel Laboratory, Functional Design Studio, Ecological and Carbon Footprint Research Laboratory and Industry Sustainability Research Laboratory.

He has supervised 40 PhD students and more than 100 research personnel. He has over 500 scientific publications, including 317 SCI/CPCI papers and 230 peer reviewed conference papers with 3783 citations and h-index of 30 and i10-index 100. He is selected on the list of "Highly Cited Researcher" in 2013 by THOMSON REUTERS, which will be used as one of the criteria for the 2013 ACADEMIC RANKING OF WORLD UNIVERSITIES.

S. S. Muthu and Y. Li, *Assessment of Environmental Impact by Grocery Shopping Bags*, EcoProduction, DOI: 10.1007/978-981-4560-20-7,

He owns more than 76 patents, including 33 patents granted in China, USA and Australia and 22 patents transferred to industry, including 16 patents and 7 software copyrights from previous ITF projects. With 80 invited keynote/plenary lectures in conferences, he has more than 49 awards on outstanding research papers, patent inventions and technology transfer, including a China National Invention Award (2nd class) and 2012 China Textile Academic Leader Award. The inventions from his team have been successfully commercialized and sold to more than 29 countries and developed as the national testing standards in USA and China (the "Moisture Management Tester") and become the authorized product by Hong Kong Hospital Authority (the "Nano Facemasks"). The high performance sportswear designed and developed by his team has been utilized to support elite athletes in major sports competitions in the world, including 2008 Beijing Olympic Game and 2012 London Olympic Game.

MIX
Papier aus verantwortungsvollen Quellen
Paper from responsible sources
FSC® C105338

If you have any concerns about our products,
you can contact us on
ProductSafety@springernature.com

In case Publisher is established outside the EU,
the EU authorized representative is:
Springer Nature Customer Service Center GmbH
Europaplatz 3, 69115 Heidelberg, Germany

Printed by Libri Plureos GmbH
in Hamburg, Germany